EFFECT OF ALGAL BIOFILM AND OPERATIONAL CONDITIONS ON NITROGEN REMOVAL IN WASTEWATER STABILIZATION PONDS

T0186482

MOHAMMED BABU

Thesis committee

Thesis supervisor
Prof. dr. H.J. Gijzen
Professor of Environmental Biotechnology
UNESCO-IHE Institute for Water Education
Delft, The Netherlands

Thesis co-supervisors
Dr. ir. N.P van der Steen
Senior Lecturer in Sanitary Engineering
UNESCO-IHE Institute for Water Education
Delft, The Netherlands

Dr. ir. C.M. Hooijmans
Senior Lecturer in Sanitary Engineering
UNESCO-IHE Institute for Water Education
Delft, The Netherlands

Other members
Prof. dr. ir. P.N.L. Lens
UNESCO-IHE Institute for Water Education
Delft, The Netherlands

Prof. dr. ir. H.H.M. Rijnaarts
Wageningen University
Wageningen, The Netherlands

Prof. dr. R. Haberl
University of Natural Resources and Life Sciences
Vienna, Austria

Prof. dr. F. Kansiime
Makerere University Institute of Environment and Natural Resources
Kampala, Uganda

This research was conducted under the auspices of the Wageningen University Institute for Environment and Climate Research (WIMEK)

Effect of Algal Biofilm and Operational Conditions on Nitrogen Removal in Wastewater Stabilization Ponds

Thesis

Submitted in fulfilment of the requirements of
the Academic Board of Wageningen University and
the Academic Board of the UNESCO-IHE Institute of Water Education
for the degree of doctor
to be defended in public
on Friday 28 January 2011 at 10.00 a.m.
in Delft, the Netherlands

by

MOHAMMED BABU
Born in Mbale, Uganda

CRC Press/Balkema is an imprint of the Taylor & Francis Group, an informal business
© 2011, Mohammed Babu

*All rights reserved. No part of this publication or the information contained herein
may be reproduced, stored in a retrieval system, or transmitted in any form or by
any means, electronic, mechanical, by photocopying, recording or otherwise,
without written prior permission from the publishers.*

*Although care is taken to ensure the integrity and quality of this publication and the
information therein, no responsibility is assumed by the publishers nor the author
for any damage to property or persons as a result of operation or use of this
publication and/or the information contained herein*

Published by:
CRC Press/Balkema
PO Box 447, 2300 AK Leiden, the Netherlands
E-mail: Pub.NL@taylorandfrancis.com
www.crcpress.com – www.taylorandfrancis.co.uk – www.ba.balkema.nl
ISBN 978-0-415-66946-7 (Taylor & Francis Group)
ISBN 978-90-8585-849-2 (Wageningen University)

Dedication

This thesis is dedicated to my late uncle Shantilal .A. Vyas

May God bless and rest your soul in peace

Acknowledgements

The author is extremely grateful to the Netherlands Government for providing financial assistance through the Netherlands Fellowship Program. I would also wish to extend my sincere thanks to the EU-SWITCH project for the financial assistance of my research project.

I'm very grateful to my promoter Professor Huub Gijzen for his kindness, guidance, valuable discussions and comments during the writing of this dissertation. I greatly appreciate his trips to Uganda and invaluable input and support in this study.

I'm particularly indebted to my co-promoters Dr. Peter van der Steen and Dr. Tineke Hooijmans for constructive ideas during the research period. I would greatly thank them for accepting to come to Uganda and providing invaluable guidance. I extend my heartfelt thanks to them for allocating time for meetings and critically reviewing the manuscripts. Special thanks go to Dr. Henk Lubberding who mentored me during my MSc study, kept in touch with me, notified and supported me when the PhD opportunity became available.

Thanks to the Managing Director and management of National Water and Sewerage Corporation for giving me permission and opportunity to use their facilities at the Bugolobi Sewage Treatment Works. I'm specifically grateful to Eng. Kiwanuka Sonko, Dr. Kaggwa Rose and Mr. Kanyesige Christopher for their assistance and link to National water.

I would wish to extend my appreciation to the Rector and Vice rector of Islamic University in Uganda, Dr. A.K. Ssengendo and Dr. M. Mpeza for giving me the opportunity and supporting me when pursuing this cause. Thanks to Dr. P.S.N.A Ssekimpi for being parental and providing guidance throughout my under and postgraduate studies. Thanks to my colleagues Dr. S. Nachuha and Mr. S. Okurut who gave me courage and support during this period.

The laboratory staff of IHE Fred, Frank, Peter, Lyzette, Don and the rest; you did a great job for me while I was in Delft. Thanks to Edwin Hes and Mushi .M. for their assistance during experimental work while in Delft. I extend my appreciation to the laboratory staff of BSTW especially Juliet, Nyombi .J. Mutyaba, C., Arra, K, Wetaga, H and Saazi Job for supporting me in the lab. Not forgetting, my cousin Rama for accepting to do the dirty work of pumping of wastewater daily into the anaerobic tank; thanks a lot dear! Thanks to my PhD colleagues Kittiwet, K., Heyddy, L., Barreto C., Uwamariya V., Sekomo, C., Bagoth S; Ansa E and rest who made me feel at home while in Delft.

Special thanks to my beloved family; my mother, father and sisters who gave me the courage and strength to move on. Great and heartfelt appreciation to my lovely wife Samiha and her family, my sons Fesal and Irfan; I deeply appreciate your patience. Thanks for tolerating my long hours in the lab and my absence from home for many months. Finally, I acknowledge all those who helped me but whose names have not appeared on this page. May God bless you all!

List of abbreviations

NH_4^+	Ammonium
NH_4^+-N	Ammonium nitrogen
et al	and others
AT	Anaerobic Tank
R_{bio}	Biofilm nitrification rate
BOD_5	Biochemical Oxygen Demand (5days)
BNR	Biological Nitrogen Removal
BSTW	Bugolobi Sewage Treatment Works
R_{bulk}	Bulk water nitrification
cm	Centimeters
COD	Chemical Oxygen Demand
CFD	Computational Fluid Dynamics
d^{-1}	per day
DO	Dissolved Oxygen
d	Dispersion number
Eff	Effluent
EU	European Union
EPS	Extra Polymeric Substance
FP	Facultative ponds
Q	Flow rate
g	Grams
hrs	Hours
HRT	Hydraulic Retention time
Inf	Influent
KjN	Kjeldahl nitrogen
MP	Maturation ponds
m	Meters
μm	Micro meters
mg	Milligrams
mg l^{-1}	Milligrams per liter
NO_3	Nitrate
NO_3-N	Nitrate nitrogen
NO_2	Nitrite
NO_2-N	Nitrite nitrogen
N_2	Nitrogen gas
N_2O	Nitrogen Oxides
Org-N	Organic nitrogen
N	Reactors in series
Pe	Peclet number
pH	Potentiometric hydrogen ion concentration
s	Seconds
SPSS	Statistical Package for Social Sciences
spp	Species
T	Temperature (^{o}C)
TSS	Total suspended Solids

UNEP	United Nations Environmental Program
US$	US dollars
WSP	Wastewater stabilization ponds

Greek

β	Standard coefficient
μE	Micro-Einstein's
α	Index of short circuiting
σ^2	Variance

Table of Contents

Curriculum vitae

Chapter 1
Introduction

Chapter 1

Introduction

Water pollution as a result of nitrogen contamination is a worldwide problem. Nitrogen is known to be detrimental to public health and the environment (Gijzen and Mulder, 2001). For instance, ammonia can cause eutrophication and oxygen depletion in the receiving water; this may result in proliferation of aquatic weeds and massive fish kills. Other effects of nitrogen pollution include underground water pollution (Laegreid et al., 1999; Zhu et al., 2005), blue-baby syndrome in infants (Bulger et al., 1989) and the emission of gasses contributing to the greenhouse effect (Takaya et al., 2003).

There are various sources of nitrogen in the environment; the major ones include agricultural runoff (Cang et al., 2004; Sheng wei et al., 2009), domestic and industrial wastewater among others. Control of non-point sources is a challenge although proper agricultural and land management practices could abate this problem (Amans and Slangen, 1994; Sheng wei et al., 2009). For point sources like domestic and industrial wastewater, pollution can be reduced through wastewater treatment.

The approach of minimizing nitrogen pollution through wastewater treatment differs between developed and developing countries. In industrialized nations, high environmental standards and stringent regulations have been set (Metcalf and Eddy, 2003) and this requires advanced wastewater treatment (Jorgensen and Williams, 2001). Activated sludge with biological nitrogen removal (BNR) is an example of advanced wastewater treatment commonly used by developed nations. This is effective in wastewater treatment but has disadvantages of high capital investments, maintenance costs and requires skilled manpower (Veenstra and Alaerts, 1996). Other systems for nutrient removal based on chemical treatment of wastewater are chemical precipitation, ammonia stripping and ion exchange. Although they are efficient, use of chemicals is undesirable and expensive (UNEP, 1999).

Developing countries cannot afford advanced wastewater treatment systems due to prohibitive costs involved in construction and maintenance (Veenstra and Alaerts, 1996). Those in tropical regions opt for natural systems which are driven by solar energy. The major systems used include constructed wetlands and the conventional wastewater stabilization ponds.

Constructed wetlands mainly use aquatic plants that have root systems which provide attachment sites for bacterial growth and activity. Alaerts et al., (1996); Bonomo et al., (1997); Gijzen and Khondker, (1997); Zimmo et al., (2000) Korner et al., (2003) and Caicedo et al., (2005) have used duckweed in wastewater treatment. The major advantages of duckweed systems are low operation costs, low energy requirements, they can withstand loading shocks and they are effective in reducing odour and TSS. They also have a potential for resource recovery (Caicedo et al., 2005). The major disadvantages include relatively large area for construction, less efficient in pathogen removal and limited nitrification. Papyrus (Okia, 2000; Kansiime and van Bruggen, 2001) and *Phragmites*

(Green and Upton, 1995; Okia, 2000) have also been used in constructed wetlands. Like in duckweed systems, nitrification is also limited in these systems due to limitation of oxygen. The wetlands also demand periodic harvesting of plant biomass for effective nutrient removal. Furthermore, the efficiency of wetlands may be reduced over time due to sedimentation.

Wastewater stabilization ponds (WSP) are the most common wastewater treatment technologies used in developing nations, especially in tropical regions. This is due to cost-effectiveness in construction and maintenance. According to UNEP (1999), wastewater stabilization ponds are still the cheapest treatment technology. They are effective in removal of organic matter (Mara and Pearson, 1998) and pathogens (Van der Steen *et al.*, 1999; Zimmo *et al.*, 2002).

The major disadvantage of wastewater stabilization ponds is the requirement of relatively large areas for construction. Large area requirement is still a big challenge and critical in application of WSP even in developing countries where land may not be expensive (Pearson, 1996). Future demographic projections indicate that by 2017, the developing world is likely to become more urban than rural in character (United Nations, 2004). Rapid urban growth of cities and towns in developing world is outstripping their capacities to provide adequate services (Cohen, 2006). This implies that as urbanization and population increases, the demand for adequate services like wastewater treatment will be higher. More space for expansion of wastewater treatment plants to achieve effective treatment will be necessary. Usually urban and population growth are associated with increased demand for land hence the cost of land is expected to become high as well (Yu *et al.*, 1997). This may be a future bottle neck to application WSP in the developing world unless research on more compact WSP systems is successful.

The other disadvantages of WSP are bad odour, mosquito breeding, high total suspended solids (TSS) in the effluent (Mara, 2004); short-circuiting (Shilton *et al.*, 2000); narrow zone for nitrification, since the aerobic zone is limited to the upper 50 cm (Baskran *et al.*, 1992), long hydraulic retention time and low nitrifier biomass in the water column (McLean *et al.*, 2000; Zimmo *et al.*, 2000).

The aim of this study was to enhance nitrogen removal in WSP. It has been proposed that introduction of attachment surfaces in wastewater stabilization ponds can improve the process of nitrogen removal (Pearson, 2005). Attachment surfaces for nitrifiers in wastewater stabilization ponds have been tried in Australia and have shown potential for application (Baskran *et al.*, 1992; Craggs *et al.*, 2000; McLean *et al.*, 2000). In this study, baffles were used as the attachment surfaces; baffles as biofilm surfaces have been tested but limited to laboratory scale studies (Kilani and Ognurombi, 1984; Muttamara and Puetpaiboon, 1997). Studies at pilot and full scale under tropical conditions were yet to be done. The effect of different arrangement of baffles on hydraulic performance and nutrient removal in wastewater stabilization ponds under tropical conditions was not known. This was one of the major issues that this research addressed. The other aspects that this research addressed include: effect of baffles on algal-bacterial biofilm structure and composition, nitrification in bulk water and biofilms; and biofilm nitrification rates under different oxygen and pH conditions.

However, before baffles are used as intervention in improving nitrogen removal in WSP (as in this research), there is need for a deeper understanding of the problems associated to nitrogen removal in WSP. The factors that affect biofilm formation should be clearly understood. Different nitrogen transformation processes in wastewater stabilization ponds were reviewed in light of the current study.

Problems associated with algae ponds
Studies by Van der Steen *et al.,* (2000a, 2000b), Zimmo *et al.,* (2000) and many others using algae in nitrogen and pathogen removal is promising. Despite the promising results, there are still a number of problems associated with wastewater stabilization ponds. These problems were categorized into two, i.e. limitation of biological processes and short-circuiting.

1. Limitation of the biological process
This includes (a) insufficient surface area for microbial attachment, (b) thermal stratification and (c) transport of wastewater from aerobic to anaerobic zone for bacterial activity.

(a) Insufficient attachment surface
Lack of surface area for nitrifiers and denitrifiers is a limitation for nitrogen removal in algae ponds (Zimmo *et al.,* 2000). Algae lack roots (as compared to other aquatic plants with long and extensive roots) hence provide a smaller surface area for microbial attachment and activity. Suspended algae and other materials in sewage can provide attachment to bacteria in the water column (McLean *et al.,* 2000). However, this is temporary attachment since suspended matter settles at the bottom of the ponds with time. Suspended algae can also be washed out of the ponds taking with them the attached bacteria. Hammer and Knight (1994) report that nitrifying bacteria live as layered outgrowths on attachment sites and rarely do they live as free floating. In this context, provision of artificial substrata becomes a good option of improving nitrifier attachment hence creating favorable conditions for their growth. The biggest challenge however is balance between nitrifiers and heterotrophic bacteria; the former are known to be slow growers and in the event of high organic loading, they can be easily out competed (Loosdrecht *et al.,* 2000).

Studies incorporating attachment surfaces in WSP have shown positive results (Baskaran *et al.,* 1992; Zhao and Wang, 1996; McLean *et al.,* 2000; Schumacher and Sekoulov, 2002). Previous work using polyvinyl acetate as the artificial attachment media has also shown improved removal efficiencies of 96%, 76% and 90% for NH_4-N, COD and BOD respectively (Zhao and Wang, 1996). McLean *et al.,* (2000) used geotextile-polyfelt TS 1600 as the carrier material and reports NH_4-N reduction from 40 mg l^{-1} in the influent to 8.7 mg l^{-1} in the effluent of algae ponds with biofilm support. It was observed that nitrifier populations in lagoons without biofilm support could only achieve high nitrification rates if

4

algal biomass was high. The algae acted as attachment substrata most especially during seasons of low wash out.

Prior to introducing biofilm surfaces for nitrifier attachment in algae ponds, deeper understanding of factors that affect biofilm development is necessary. Development of attached microbial growth is a very complex process involving many variables. The major ones include oxygen, pH, temperature, nutrients, cations, substrata, extracellular polymeric substance (EPS) and flow velocity (Characklis et al., 1990; Esterl et al., 2003).

Oxygen is vital in respiration for all aerobic living organisms. It serves as the final electron acceptor in the electron carrier system during provision of energy. For any metabolic activity to be sustained, energy requirement must be fulfilled. In practical terms, bulk water dissolved oxygen concentration of 2.0 to 3.0 mg l^{-1} is satisfactory for aerobic suspended nitrifier growth (Grady et al., 1999; Metcalf and Eddy, 2003; Arceivala and Asolekar, 2008). For attached growth systems, this may be higher especially with mass transfer limitations (Loosdrecht et al., 2000). Oxygen limitation is one of the main factors responsible for biofilm sloughing. During biofilm development, thicker slime layers are laid on the existing ones. Substrate and oxygen is consumed from the surrounding wastewater before penetrating deeper. The bacteria in the deeper region undergo endogenous respiratory state and lose the ability to cling on the surface (Metcalf and Eddy, 2003). This results into biofilm sloughing.

Wastewater has a wide range of pH variations making it suitable for diverse growth of microorganisms. Nitrifiers and denitrifiers have been reported to grow well in pH ranges from 7.2 to 9.0 and 7.0 to 8.0 respectively (Metcalf and Eddy, 2003). Very high or very low pH values are detrimental to physiology of microorganisms. Schumacher and Sekoulov (2002) observed increased pond pH by algal biofilms (due to consumption of carbon dioxide by the photosynthesis). The resulting high pH values caused a decrease in nitrogen removal (Table 1.1). This is an indication that probably the process of denitrification was affected; which was to the disadvantage of the treatment process. Previous studies (Caicedo et al., 2005) have shown that combination of algae and duckweed systems can counteract the effect of high pH. In duckweed ponds, light penetration is limited hence algal productivity is virtually absent. Stable pH of 6.8 to 7.0 and relatively low oxygen levels has been reported in these ponds. Contrary to Schumacher and Sekoulov (2002), it is thought that biofilms provide a variety of microenvironments that could favor treatment processes. For instance, the outer aerobic zones of the algal biofilms are ideal for nitrifiers while the deeper anaerobic zones could be favorable to denitrifiers.

Table 1.1 Effect of bulk water pH on nitrogen removal rates by nitrification-denitrification (Schumacher and Sekoulov, 2002)

pH	Removal rates (g-N $m^{-2}hr^{-1}$)
7.0-8.0	0.36
8.0-9.0	0.19
9.0-10.0	0.16
10.0-11.0	0.097

Temperature is a driving force for many metabolic reactions. Very high or low temperatures tend to affect enzyme activity thus limiting metabolic reactions. Studies by Donlan et al., (1994) on effects of seasonal variations on biofilm formation in drinking water caste iron pipes indicated increased growth with increased temperature. In their work, all the study sites sampled under warm conditions (15 to 25^{0}C) had higher rates of biofilm formation compared to those under cold conditions (4 to 15^{0}C). This is important for tropical and subtropical regions where temperature is normally high. It can be expected that biofilm formation in these regions will be higher and probably provide diverse microenvironments for effective wastewater treatment.

Nutrients in wastewater are abundant in the bulk water; hence limitation mostly occurs in biofilms. Different ratios of nutrients available determine the type of biofilm that develops (Loosdrecht et al., 2000). Higher concentration of biodegradable organic substances usually favors the growth of heterotrophs. Microorganisms consume nutrients in their vicinity creating a nutrient gradient. The gradient causes nutrient replenishment, which is advantageous to the fast growers normally found on biofilm surfaces. The slow growers are usually relegated to the base (Loosdrecht et al., 2000) and due to mass transfer limitations; they may enter endogenous respiratory state and be sloughed (Lewandoski and Beyenal, 2003). Increase in nutrients results in increased growth (Cowan et al., 1991) as long as other factors do not become limiting.

Cations are a vital component in biofilm development. They increase bacteria attachment to surfaces by either physiology-dependant mechanisms or by reducing negative repulsive forces between bacteria and surfaces. Electrolytes such as calcium and magnesium are important cellular cations and cofactors for enzymatic reactions. These play a role of enhancing attachment indirectly. Alternatively, cations improve bacterial attachment by reducing repulsive forces between negatively charged bacterial cells and glass surfaces through neutralizing the charges. Fletcher, (1988) was able to demonstrate this phenomenon in experiments on the effects of Sodium, Calcium, Lanthanum and Iron (III) on attachment of *Pseudomonas fluorescens* to glass. It was found that the cations reduced repulsive forces between any two groups of adhesive polymers found in slimes.

One of the most important factors for attached growth is the surface characteristic. Rough and porous surfaces have been found to be suitable for microbial growth. These provide increased surface area for attachment and protection against hydraulic shear forces (Oliveira et al., 2003). A diversity of microorganisms will colonize rough and porous surfaces more rapidly due to the variety of microenvironments created (Characklis et al., 1990).

Substrata with higher hydrophobicity or wettability are known to favor microbial attachment. Hydrophilic substances attract water, apparently bringing the cells closer to the substratum. Hydrophilic interactions can also prevail between the cells and the substratum thereby reducing repulsive forces (Oliveira et al., 2003). Non-polar substances like Teflon and plastics attach microorganisms more rapidly than glass and metals.

Attachment cannot be possible without a matrix upon which cells are deposited. This is provided by a substance known as extracellular polymeric substance (EPS) made by the bacteria. EPS is composed of nucleic acids, proteins and other organic matter (85-90%). Its development depends on the nutritional status of the surrounding media. EPS are highly hydrated substances due to the hydrogen bond formation with water. This property enables them to prevent cells from desiccation. EPS are also known to protect microorganisms against toxic substances and anti-biotics (Bishop, 2003).

EPS have unique properties of possessing negatively charged groups on their surface. This permits binding to cations such as calcium and magnesium, which are known to form cross-links with polymer strands providing greater binding force in biofilms (Fletcher, 1988; Bishop, 2003). EPS are essential in wastewater treatment, variations in its biological, chemical and physical properties make treatment technologies like activated sludge, trickling filters, rotating biological contactors, fluidized or submerged fixed-bed reactors depend on them (Bryers and Characklis 1990; Metcalf and Eddy, 2003).

For organisms to attach, the rate of attachment should be greater than the washout rate. These two processes are greatly influenced by velocity. The zone adjacent to the substratum-liquid interface is termed as the hydrodynamic boundary layer. Its thickness depends on linear velocity; the higher the velocity, the thinner it becomes. Increasing flow velocity exerts mechanical stress on the biofilm thus wearing it out (Esterl et al., 2003). Although beyond a certain threshold it may erode and abraise the biofilm (Morgenroth, 2003), some degree of velocity may have a positive effect. Rijnaarts et al., (1993) and Donlan et al., (1994) have shown that fluid movement aids transportation of cells to the substratum for deposition.

(b) Thermal stratification
Temperature differences can greatly influence effluent quality. Thermal stratification leads to formation of water layers, each with different characteristics of temperature, oxygen, pH and redox potential. This effects denitrification since movement of nitrates from upper aerobic to deeper anaerobic water layers may not occur. Stratification can also strongly influence removal of pathogens from wastewater through photo oxidation (Curtis et al., 1992, 1994). Prevention of mixing more or less limits photo oxidation to top layers. Several authors (Pescod and Almansi 1996; Van der Steen et al., 2000a) have observed this effect in wastewater stabilization ponds. Shilton and Harrison (2003) recommend overcoming this problem by adequate mixing of the influent into the main stream.

(c) Transport of wastewater from aerobic to anaerobic zones
In open water, aerobic zones are known to exist in the upper layers where there is sufficient light and contact with atmospheric oxygen. These zones are mostly created by photosynthetic evolution of oxygen by the algae; this is documented to be the basis of a

successful wastewater treatment in oxidation ponds (Mara *et al.*, 1992; Fruend *et al.*, 1993). The aerobic zones tend to favor the process of nitrification. Caicedo, (2005) observed nitrification in upper water layers of duckweed ponds in which open aerobic zones were created. The deeper water layers are usually anaerobic due to insufficient light to promote photosynthesis. The oxygen produced in the open upper layers is usually consumed within the vicinity and little of it diffuses to the deeper water layers. The deeper zones are thus anoxic providing the right conditions for denitrification. In horizontal flow systems, the nitrates formed in the aerobic zones are transported to lower anoxic zones (where denitrification occurs) by diffusion. This movement may not be effective especially in the event of thermal stratification. For effective nitrification, a mechanism of movement of nitrates from the aerobic to anaerobic zone may be necessary. Nitrates can be moved to the deeper zones by the alternating upward and downward flow patterns induced by the vertical baffles.

2. Short- circuiting

Short-circuiting can cause significant loss of treatment efficiency. A small deviation from the anticipated hydraulic retention time (HRT) can lead to poor effluent quality (Shilton *et al.*, 2000). According to Shilton and Harrison, (2003) and Barbagallo *et al.*, (2003), the major causes of short-circuiting in wastewater stabilization ponds are:

(a) Inlet momentum
(b) Wind effects
(c) Temperature effects

(a) In-let momentum

This is related to inlet types. Shilton and Harrison (2003) have analyzed different inlet types and have pointed out the shortcomings of each of them. Table 1.2 summarizes this information.

Table 1.2 Different types of inlets and their shortcoming (Shilton and Harrison, 2003)

Inlet type	Remarks
Horizontal inlet (Horizontal pipe submerged in the water)	Jetting effect drives the bulk volume into circulation patterns with higher velocity than inflow. Influent moves faster to outlet
Large horizontal inlet (Larger pipes)	Same effect as horizontal inlet but only delays short-circuiting
Vertical inlet	Fluid moves into two plumes along the two adjacent walls.
Vertical inlet with stub baffles	Improved flow but limitation of overloading in the inlet zone due to the stubs
Diffuse inlet (Manifold)	Improves flow, expensive to install and maintain, recommended for pre-treated water
Inflow dropping from horizontal pipe (Horizontal pipe discharging at top of water surface)	Has the same effect as the horizontal inlet despite the fact that water drops vertical into the bulk.

The inlet type depends on the pond type. For instance in ponds that receive high organic or solid loading, a horizontal inlet pipe ensures good mixing of the influent into the pond mass. However, to prevent the influent from swirling quickly to the outlet, use of stub baffles is recommended. For ponds that receive pre-treated influent (e.g. maturation ponds), diffuse inlets are recommended; although these are expensive to install in full scale systems (Shilton and Harrison, 2003).

(b) Wind effects

Wind effect is mainly experienced in areas or seasons where wind velocity is high. The wind currents force the influent to move rapidly to the outlet resulting into short- circuiting. Although there are suggestions that wind currents add oxygen to the water, its contribution has been found insignificant. Algae play a more important role in oxygenation than wind (Shilton and Harrison, 2003).

(c) Temperature effects

Temperature differences cause thermal stratification that result in vertical density boundaries. This prevents vertical mixing thus inflow can short-cut at the top layer of the stratified pond. This may significantly affect treatment. Therefore a system designed to break up stratification is desired.

Microbiology
Nitrogen transformations in wastewater treatment

They are various forms of nitrogen in the aquatic environment. The major nitrogen transformation routes in wastewater stabilization ponds are shown in figure 1.1. The transformations include ammonification, nitrification, denitrification, ammonia volatilization, assimilation, sedimentation, fixation and anaerobic ammonia oxidation (ANNAMOX).

Ammonification

This is the biological transformation of organic nitrogen to ammonia. It is a microbial mediated process which is carried out by the ammonifying bacteria through mineralization of amino acids, urea and uric acid to ammonia. During this process, energy is released which the bacteria utilize for metabolic activities. Ammonifying bacteria can also use ammonia directly in building biomass. Transformation of organic nitrogen to ammonia normally proceeds at a faster rate under sufficient oxygen concentrations. It is usually faster than nitrification and if the latter is affected, ammonia accumulation occurs. This happens mostly during a rapid change from aerobic to anaerobic conditions. The optimum temperature and pH for ammonification range from $40^{o}C$ to $60^{o}C$ and 6.5 to 8.5, respectively. These temperature ranges rarely occur in biological treatment systems such as algae ponds and wetlands (Kadlec and Knight, 1996). In conventional treatment systems, ammonification is usually considered a first order kinetic reaction. For wetland systems, organic nitrogen decreases with time in agreement with first order kinetics (Kadlec and Knight, 1996).

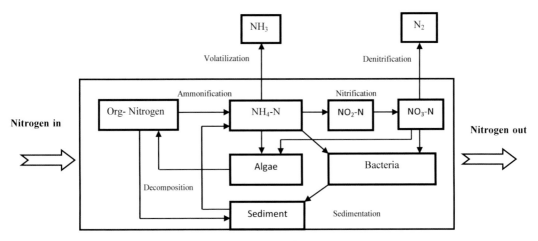

Figure 1.1: Major nitrogen transformation routes in algae ponds

Nitrification

This is an important biological process in wastewater treatment which occurs by two-step oxidation of ammonia. There are a number of autotrophic nitrifying bacteria that perform nitrification but the most important genera are *Nitrosomonas* and *Nitrobacter* (Metcalf and Eddy, 2003). Ammonia oxidation is an aerobic process that requires oxygen. In *Nitrosomonas spp* for instance, oxygen is used by the enzyme ammonium oxygenase that initiates the ammonia oxidation pathway (Tiedje, 1988). The two-step reaction starts with oxidation of ammonia to nitrite by *Nitrosomonas spp* followed by conversion of the nitrites to nitrates by *Nitrobacter spp*.

$$2NH_4^+ + 3O_2 \longrightarrow 2NO_2^- + 4H^+ + 2H_2O \qquad (1)$$
$$2NO_2^- + O_2 \longrightarrow 2NO_3^- \qquad (2)$$

The nitrates formed can be assimilated by other organisms or can be denitrified to dinitrogen gas. Apart from denitrification, nitrates can also be transformed to other nitrogen forms through two other processes namely; assimilatory nitrate reduction and dissimilatory nitrate reduction. Assimilatory nitrate reduction involves the reduction of nitrate to ammonia by bacteria (Metcalf and Eddy, 2003). The ammonia produced is usually used up in cell synthesis especially in absence of NH_4^+ ions in the growth medium. Attached algal biofilms have also been reported to carry out assimilatory nitrate reduction on their cell surfaces under high pH (>10) and oxygen concentrations of 9 mg l^{-1} (Schumacher and Sekoulov, 2002). Assimilatory nitrate reduction is mainly regulated by ammonium and carbon. On the other hand, dissimilatory nitrate reduction also involves the reduction of nitrate to ammonia but the process is regulated by oxygen and it is unaffected by ammonium. This pathway is well suited for anaerobic environments (Tiedje, 1988). In dissimilatory nitrate reduction, nitrate reduction to ammonia is for the purpose of energy yield rather than for biomass development. This implies that the ammonium produced would accumulate in the surrounding medium. Bulger *et al.*, (1989) found this mechanism to account for higher ammonium concentration in ground water polluted with nitrate rich leachate from a land fill and seepage from wastewater stabilization ponds. The oxidizing ground water from the land fill provided the source of nitrates while the seepage from the waste stabilization ponds provided organic matter and reducing conditions. This favored dissimilatory nitrate reduction hence increasing the ammonium concentrations in the ground water to levels higher than that found in the lagoons.

For nitrification to occur, sufficient oxygen must be present. Metcalf and Eddy (2003) reported 4.57g O_2 and 7.14g of alkalinity (calcium carbonate) as a requirement for complete oxidation of 1g of NH_4^+ - N. Dissolved oxygen less than 0.50 mg l^{-1} is thought to limit nitrification in suspended growth under laboratory conditions. Other factors that affect nitrification include temperature, pH, BOD, toxic compounds and high concentration of other forms of nitrogen (Arceivala and Asolekar, 2008). For pure bacterial cultures, temperature range from 25^0C to 35^0C has been found to be optimum for nitrification (Kadlec and Knight, 1996). The optimum pH values required in suspended growth range from 7.2 to 9.0 (Metcalf and Eddy, 1991). Lower pH values of 6.5 have also been reported for pure cultures of ammonia oxidizers (Princic et al., 1998). High BOD levels are known to favor growth of heterotrophic bacteria. Since nitrifiers are slow growers, they are usually out-competed by heterotrophic bacteria under conditions of high BOD. Other forms of nitrogen that inhibit nitrification include free ammonia, nitrous acid and nitrites. It has been found that concentrations of 100 mg l^{-1} of NH_4^+ - N and 20 mg l^{-1} of NO_2 -N at pH 7.0 and temperature of 20^0C inhibits ammonia and nitrite oxidation (Metcalf and Eddy, 2003).

The kinetics of nitrification follows the Monod equation. It is an aerobic process, which can be controlled by dissolved oxygen flux from the atmosphere. Usually, oxygen flux into wetlands and treatment systems is by mass transfer from the atmospheric sources and this is a first order process (Kadlec and Knight, 1996).

Denitrification
This is the biological reduction of nitrate to nitric oxide, nitrous oxide, and nitrogen gas by microorganisms. It is a vital process in wastewater treatment where prevention of eutrophication and NO_3-N pollution of ground water is required (Metcalf and Eddy, 2003). In nature, denitrification is an important process because it closes the loop of the nitrogen cycle. Without this process, atmospheric nitrogen would be depleted (Kadlec and Knight, 1996).

Denitrification is favored in absence of oxygen (anoxic or anaerobic) although most denitrifiers are facultative. In absence of dissolved oxygen, the denitrifying bacteria use the oxygen bound to the nitrate or nitrite as the final electron acceptor (Kadlec and Knight, 1996). Under high oxygen levels, most common denitrifiers use oxygen as the final electron acceptor in preference to nitrate; this is due to the high energy yields. Furthermore, the denitrifiers require high amounts of energy for splitting the nitrogen-oxygen strong bonds in nitrate; they usually tend to avoid this path. Denitrification in the bulk water will cease under high oxygen conditions but the process still continues to occur in the microscopic anoxic zones of biofilms.

Another important factor known to limit denitrification is COD since absence of carbon inhibits denitrification. It is estimated that 2.86g of COD is required to denitrify 1 g of NO_3-N (Oostrom, 1995). Utilization of carbon by denitrifiers is coupled with production of alkalinity. For instance, for every 1g of NO_3-N reduced with methanol, 3.0 g of bicarbonate as $CaCO_3$ is produced (Kadlec and Knight, 1996). Increased production of alkalinity causes increase in pH and high values are detrimental to many microorganisms. The pH and temperature ranges suitable for denitrifier growth is from 7 to 8 and 5^0C to 25^0C respectively.

Denitrification is not limited to only anaerobic conditions; recent studies have shown that aerobic denitrifiers such as *Paracoccus denitrificans* (formerly *Thiosphaera pantotropa*), *Microvirgula aerodenitrificans* and *Thaurea mechernichesis* exist. *Paracoccus denitrificans* is known to reduce nitrates even at oxygen saturation levels (Loyd *et al.*, 1987; Robertson and Kuenen, 1990; Takaya *et al.*, 2003).

Although denitrification is a key nitrogen removal mechanism in wastewater (Zimmo *et al.*, 2000; Metcalf and Eddy, 2003; Caicedo, 2005), denitrifiers (both aerobic and anaerobic) produce more nitrous oxide than nitrogen gas (Takaya *et al.*, 2003). Investigations on the activities of *Pseudomonas stutzeri* and *Paracoccus denitrificans* under different oxygen concentrations (anoxic, hyperoxic and oxic) have been done (Takaya *et al.*, 2003). Results indicated that *Pseudomonas stutzeri* in particular, does not produce nitrogen gas under anoxic conditions. It is argued that complete anoxic conditions barely exist or exist to a small extent in treatment systems that use aerobic nitrification followed by anaerobic denitrification. Under such conditions, these bacteria have a high potential of nitrous oxide production compared to nitrogen gas production. Same studies have also shown that the typical aerobic denitrifiers, *Paracoccus denitrificans* produce more N_2O than nitrogen gas under oxic conditions. Two strains of aerobic denitrifiers, *Pseudomonas stutzeri* TR2 and *Pseudomonas sp.* strain K50 are able to produce more nitrogen gas and less N_2O under aerobic conditions (Takaya *et al.*, 2003).

Ammonia Volatilization
This is the loss of un-ionized ammonia to the atmosphere. Ammonia is volatile and can be lost to the atmosphere through diffusion at conditions of high pH and temperature. In water, ammonia exists as un-ionized (NH_3) and ionized (NH_4^+) forms, and existence of the two species is pH and temperature dependent. At high pH and temperature, the ammonia fraction dominates over ammonium. The percentage of un-ionized ammonia in relation to pH and temperature can be determined using the equation proposed by Emerson *et al.*, (1975) and Pano and Middlebrooks (1982):

$$\% \text{ Un-ionized } NH_3 = 100/ (1+10^{(pKa-pH)}) \qquad (3)$$

$$pKa = 0.09108 + (2729.92/ 273.2 + T)$$

Where T = temperature ^0C

Although algae ponds are known to have elevated pH values, ammonia volatilization has been found to be an insignificant removal mechanism in these ponds (Zimmo *et al.*, 2003).

Nitrogen fixation

This is a biological process where atmospheric nitrogen is reduced to ammonia nitrogen by bacteria and cyanobacteria. It is an adaptive process for organisms living in nitrogen deficient conditions. It rarely occurs in nitrogen rich environments such as wastewater (Kadlec and Knight, 1996). This is due to repression of the process by presence of ammonia under such conditions. This transformation is negligible in wastewater treatment systems.

Assimilation and decomposition

This is the conversion of inorganic nitrogen into organic forms for cell and tissue synthesis in living organisms. In plants, ammonia nitrogen is much preferred to nitrate nitrogen. Nitrate uptake may be favored in ammonia-deficient, nitrate-rich conditions. In wastewater stabilization ponds, death and decomposition of phytoplankton may partly release the assimilated nitrogen into the water. This provides a mechanism of internal cycling of nutrients.

ANNAMOX

ANNAMOX is a process discovered by Mulder *et al.*, (1995). It is an anaerobic biological process in which ammonia is converted to nitrogen gas with nitrite as the electron acceptor. Autotrophic bacteria of the order *Planctomycetes* are known to carry out this process. It does not require a carbon source as compared to denitrification (Dongen *et al.*, 2001) and requires low oxygen concentration. Thus, it may be a substituting mechanism for denitrification under conditions of limited availability of organic matter.

Scope of this Thesis

This thesis presents work done on nitrogen removal both in the laboratory and in pilot scale wastewater stabilization ponds. Laboratory work included studying biofilm and bulk water nitrification rates under different environmental conditions. The rates were then used in the nitrogen mass balances of the pilot scale wastewater stabilization ponds. The ponds were constructed and operated under tropical conditions in Kampala - Uganda. There were four pilot scale ponds; with three of them fitted with baffles. It is known that nitrogen removal in wastewater stabilization ponds is limited by nitrification process which in turn is limited by lack of attachment surface for nitrifiers. The baffles in this case provided more surface area for biofilm attachment required for nitrifier growth. The pilot scale system was operated under two operational conditions i.e. under low ammonia loading (Period 1) and high ammonia loading (Period 2). The results presented in this thesis are divided according to the two operational conditions. This research was limited to only nitrogen removal in wastewater stabilization ponds. Other aspects such as BOD, COD, TSS, pH, oxygen, temperature etc. were considered only as monitoring parameters and were used in explaining nitrogen removal processes.

The first chapter of this thesis is the introduction which discusses various aspects of wastewater stabilization ponds with respect to nitrogen removal. Different paths of nitrogen transformation in wastewater stabilization ponds have been addressed here.

Chapter **2** mainly focused on biofilm characteristics i.e. biofilm growth rate, dry and wet biomass as well as algal and zooplankton species composition in wastewater stabilization ponds. These are important factors that could influence processes in wastewater stabilization ponds. In chapter **3**, results of tracer studies are presented. The effect of baffles on pond hydraulic characteristics was investigated. Chapter **4** analyzed nitrification rates in bulk water and biofilm. Laboratory activity tests were performed to discriminate between bulk water and biofilm nitrification rates; their relative importance in nitrogen removal was discussed.

The effect of light, dark, oxygen and pH on biofilm nitrification rates are described in chapter **5**. Chapter **6** presents the results of the general performance of the four ponds. The effects of operational conditions and baffles on nitrogen removal in the ponds were studied here. Chapter **7** focused on nitrogen mass balances; bulk water and biofilm nitrification rates of the pilot scale ponds were determined. These were fitted in the Kjeldahl nitrogen mass balance equation and used to predict effluent Kjeldahl nitrogen of the ponds. Total nitrogen mass balances were also performed in chapter **7**. Finally, the thesis ends with a general summary of the study and outlook.

References
Alaerts, G.J., Rahman, M.M. and Kelderman, P. (1996). Performance analysis of a full-scale duckweed covered sewage lagoon. *Water Resources Development*, **30** (4); 843-852

Amans, E.B. and Slangen, J.H.G. (1994). The effect of controlled released fertilizer 'Osmocote' on plant growth, yield and composition of onion plants. *Fertilizer Research*, **37**; 79-84

Arceivala, S.J. and Asolekar, S.R. (2008). Wastewater Treatment for Pollution Control and Re-use, Third edition, McGraw-Hill Publishers, New Delhi; p 140

Barbagallo, S., Brissaud, F., Cirelli, G.L., Consoli, S., and Xu, P. (2003). Modeling of bacterial removal in wastewater storage reservoir for irrigation purposes: a case study in Sicily, Italy. *Wat. Sci. Tech. 3 (4); 169-175*

Baskaran, K., Scott, P.H. and Connor, M.A. (1992). Biofilms as an Aid to Nitrogen Removal in Sewage Treatment Lagoons. *Wat. Sci. Tech. 26 (7-8), 1707-1716*

Bishop, P.L. (2003). The effect of biofilm heterogeneity on metabolic processes. In: Biofilms in Wastewater Treatment, An interdisciplinary Approach, P. Bishop, S. Wuertz and P. Wilderer (Ed). IWA publishing House. UK; 125-146

Bonomo, L., Pastorelli, G. and Zambon, N. (1997). Advantages and limitations of duckweed-based wastewater treatment systems. *Wat. Sci. Tech. 35 (5); 239-146*

Bryers, J.D. and Characklis, W.G. (1990). Biofilms in water and wastewater treatment. In: Biofilms, Characklis, W.G., Marshall, K.C, (Ed.) John Wiley & Sons, New York; 671-696

Bulger, P.R., Kehew, A.E. and Nelson, R.A. (1989). Dissimilatory Nitrate Reduction in a Waste Water Contaminated Aquifer. *GROUND WATER, 27* (5); September-October, 664-671

Caicedo J.R, van der Steen, NP, Gijzen HJ (2005). The effect of anaerobic pre-treatment on the performance of duckweed stabilization ponds. IWA Wat. Env. Man. Series 11, pp. 64-78

Caicedo, J.R. (2005). Comparison of performance of full-scale duckweed and algae stabilization ponds. In: Effect of operational variables on nitrogen transformations in duckweed stabilization ponds. PhD Thesis, Wageningen University and UNESCO-IHE Institute for Water Education, Delft- The Netherlands; 129-145

Cang, H.J., Xu, L.F., Li, Z. and Ren, H. (2004). Nitrogen losses from farmland and agricultural non point source pollution. *Tropical geography 25; 332-336*

Characklis, W.G., McFeters, G.A. and Marshall, K.C. (1990). Physiology ecology in biofilms. In: Biofilms, Characklis, W.G., Marshall, K.C, (Ed). John Wiley & Sons, New York; 341-394

Cohen, B. (2006). Urbanization in developing countries: Current trends, future projections, and key challenges for sustainability. *Technology in society 28; 63-80*

Cowan, M.M., Warren, T.M. and Fletcher, M. (1991). Mixed species colonization of solid surfaces in laboratory biofilms. *Biofouling 3; 23-34*

Craggs, L.J., Tanner, C.C., Sukias, J.P.S., Davies, C.R.J. (2000). Nitrification potential of attached biofilms in dairy wastewater stabilization ponds. *Wat. Sci. Tech. 42(10-11), 195-202*

Curtis, T.P., Mara, D.D. and Silva, A.S. (1992a). Influence of pH, Oxygen, and Humic substances on the ability of sunlight to damage fecal coliforms in waste stabilization pond water. *Appl. Env. Microbiol. 58 (4); 1335-1343*

Curtis, T.P., Mara, D.D., Dixo, N.G.H. and Silva, A.S. (1994). Light penetration in waste stabilization ponds. *Wat. Res. 28(5); 1031-1038*

Dongen, L.G.J.M van., Jetten, M.S.M. and Loosdrecht, M.C.M. (2001). The Combined Sharon/Anammox process, a sustainable method for nitrogen removal from sludge water, Stowa, IWA publishing UK; p 57

Donlan, R.M., Pipes, W.O. and Yohe, T.L. (1994). Biofilm formation on cast iron substrata in water distribution systems. *Wat. Res. 28 (6); 1497- 1503*

Emerson, K., Russo, R.E., Lund, R.E. and Thurston, R.V. (1975). Aqueous ammonia equilibrium calculations: Effect of pH and Temperature. *Journ. Fish. Res. Board of Canada 32 (12) 2379-2383*

Esterl, S., Hartmann, C. and Delgado, A. (2003). On the influence of fluid flow in a packed-bed biofilm reactor. In: Biofilms in Wastewater Treatment, An interdisciplinary Approach, P. Bishop, S. Wuertz and P. Wilderer (Ed). IWA publishing House. UK; 88-116

Fletcher, M. (1988). Attachment of *Pseudomonas fluorecens* to Glass and Influence of Electrolytes on Bacterium-Substratum Separation distance. *Journ of Bacteriology 170 (5); 2027-2030*

Fruend, C., Romem, E., and Post, A.F. (1993). Ecological physiology of an assembly of photosynthetic micro algae in wastewater oxidation ponds. *Wat. Sci. Tech. 27 (7-8); 143-149*

Gijzen, H.J. and Khondker, M. (1997). An overview of ecology, physiology, cultivation and application of duckweed, Literature review. Report of Duckweed Research project. Dhaka, Bangladesh.

Gijzen, H.J. and Mulder, A. (2001). The global nitrogen cycle out of balance. *Water 21*, Aug 2001; 38-40

Grady, C.P.L., Daigger, J.G.T. and Lim, H.C. (1999). Multiple microbial activities in a single continuous stirred tank reactor. In: Biological Wastewater Treatment, Second edition. Published by Marcel Dekker Inc, New York; 191-211

Green, M.B. and Upton, J. (1995). Constructed reed beds: Appropriate technology for small communities. *Wat. Sci. Tech. 32 (3); 339-348*

Hammer, D.A. and Knight, R.L. (1994). Designing constructed wetlands for nitrogen removal. *Wat. Sci. Tech. 29(4), 15-27*

Jorgensen, S.E. and Williams, D.W. (2001). Water quality; the impact of eutrophication. In: Lakes and Reservoirs, volume 3. UNEP-International Environment Technology Centre

Kadlec, R.H. and Knight, R.L. (1996). Treatment wetlands. CRC - Lewis publishers'; 373-442

Kansiime, F. and van Bruggen J.J. (2001). Distribution and retention of fecal coliforms in Nakivubo wetland in Kampala, Uganda. *Wat. Sci. Tech. 44 (11-12); 199-206*

Kilani, J.S. and Ognurombi. J.A. (1984). Effects of baffles on the performance of model wastewater stabilization ponds. *Wat. Res. 18(8) 941-944*

Korner, S., Vermaat, J.E. and Veenstra, S. (2003). Reviews and Analyses- The Capacity of Duckweed to Treat Wastewater: Ecological considerations for Sound Design. *J. Environ. Qual. Vol. 32 Sept- Oct; 1583-1590*

Laegreid, M., Bockman, O.C. and Kaarstad, O. (1999). Agriculture, Fertilizers and Environment. CAB international. Wallingford and Norsk Hydro, ASA, Oslo

Lewandowsiky, Z. and Beyenal, H. (2003). Mass transport in heterogeneous biofilms. In: Biofilms in Wastewater Treatment, An interdisciplinary Approach, P. Bishop, S. Wuertz and P. Wilderer (Ed). IWA publishing House. UK; 145-176

Loosdrecht, M.C.M., Benthum, W.A.J. and Heijnen, J.J. (2000). Integration of nitrification and denitrification in biofilm airlift suspension reactors. *Wat. Sci. Tech. 41(4-5), 97-103*

Loyd, D.; Boddy, L. and Davies, K.J.P. (1987). Persistence of bacterial denitrification capacity under aerobic conditions: the rule rather than expectation. *FEMS Microbiol. Ecol 45; 185-190*

Mara, D.D. (2004). Domestic wastewater treatment in developing countries. Earth scan, London

Mara, D.D. and Pearson, H.W. (1998). Design manual for waste stabilization ponds in Mediterranean countries. European Investment bank. Lagoon Technology International Ltd Leeds, England

Mara, D.D., Alabster, G.P., Pearson, H.W. and Mills, S.W. (1992). Waste stabilization ponds, a design manual for Eastern Africa, Lagoon Technology International Leeds, England

McLean, B.M., Baskran, K. and Connor, M.A. (2000). The use of algal-bacterial biofilms to enhance nitrification rates in lagoons: Experience under laboratory and pilot scale conditions. *Wat. Sci. Tech. 42 (10-11); 187-194*

Metcalf and Eddy (1991). Wastewater engineering. Treatment, Disposal and Reuse, 2nd Ed. Revised by Tchobanoglous, G., Burton, F.L. McGraw Hill, Inc., USA
Metcalf and Eddy, (2003). Wastewater engineering. Treatment and Reuse. Tchobanoglous, G., Burton, F.L., Stensel, H.D (Eds). 4th Ed. McGraw Hill, Inc., USA

Morgenroth, E. (2003). Detachment: an often-overlooked phenomenon in biofilm research and modeling. In: Biofilms in Wastewater Treatment, An interdisciplinary Approach, P. Bishop, S. Wuertz and P. Wilderer (Ed). IWA publishing House. UK; 264-293

Mulder, A., van der Graaf, A.A., Robertson, L.A., Uenen, J.G. (1995). Anaerobic ammonia oxidation discovered in denitrifying fluidized bed reactor. *FEMS. Microbiol. Ecol. 16; 177-183*

Muttamara, S. and Puetpaiboon, U. (1997). Roles of Baffles in Waste Stabilization Ponds. *Wat. Sci. Tech. 35(8) 275-284*

Okia, T.O. (2000). A pilot study on Municipal Wastewater Treatment Using Constructed Wetlands in Uganda. PhD Dissertation, Wageningen University and UNESCO - IHE Institute for Water Education, Delft- The Netherlands

Oliveira, R., Azeredo, J. and Teixeira, P. (2003). The importance of physicochemical properties in biofilm formation and activity. In Biofilms in Wastewater Treatment, An interdisciplinary Approach, P. Bishop, S. Wuertz and P. Wilderer (Ed). IWA publishing House. UK

Oostrom, A.J. van (1995). Nitrogen removal in constructed wetlands treating nitrified meat processing effluent. *Wat. Sci. Tech. 32 (3); 137-147*

Pano, A. and Middlebrooks, E.J. (1982). Ammonia nitrogen removal in facultative wastewater stabilization ponds. *Journ. Wat. Poll. Cont. Fed. 54 (4); 344- 351*

Pearson, H. (2005). Microbiology of waste stabilization ponds. In: Pond Treatment Technology, A. Shilton (Ed). IWA publishing, London; 14-48

Pearson, H.W. (1996). Expanding the horizons of pond technology and application in an environmentally conscious world. *Wat. Sci. Tech. 33 (7); 1-9*

Pescod, M.B. and Almansi, A. (1996). Pathogen removal mechanisms in anoxic wastewater stabilization ponds. *Wat. Sci. Tech. 33 (7); 133-140*

Princic, A., Mahne, I., Megusar, F., Eldor, P.A. and Tiedje, J.M. (1998). Effects of pH and oxygen and ammonium concentrations on community structure of nitrifying bacteria from wastewater. *Appl. Environ. Microbiol. 64 (10) 3584-3590*

Rijnaarts, H.H.M., Norde, W., Bouwer, E.J., Lyklema, J., and Zehnder, A.B.J. (1993). Bacterial Adhesion under Static and Dynamic Conditions. *Appl. Environ. Microbiol. 59 (10); 3255-3265*

Robertson, L.A. and Kuenen, J.G. (1990). Physiological and Ecological aspects of aerobic denitrification, a link with heterotrophic nitrification? In: Denitrification in soils and sediments, Edited by N. P. Revsbech and J. Serensen, Plenum Press, New York, 1990

Schumacher, G. and Sekoulov, I. (2002). Polishing of secondary effluent by an algal biofilm. *Wat. Sci. Tech. 46 (8); 83-90*

Sheng-wei, N., Wang-seng, G., Yuan-quan, C., Peng, S. and Eneji, A.E. (2009). Review of current status and research approaches to nitrogen pollution in farmlands. *Agricultural Sciences in China* **8** *(7); 843-849*

Shilton, A. and Harrison, J. (2003). Guidelines for the Hydraulic design of waste stabilization ponds, Institute of technology and engineering, Massey University, New Zealand

Shilton, A., Wilks, T., Smyth, J. and Bickers, P. (2000). Tracer studies on a New Zealand waste stabilization pond and analysis of treatment efficiency. *Wat. Sci. Tech.* **42** *(10-11); 343-348*

Takaya, N., Catalan-sakairi, M.A.B., Sakaguchi, Y., Kato, I., Zhou, Z. and Shoun, H. (2003). Aerobic denitrifying bacteria that produce low levels of nitrous oxide. *Appl. Environ. Microbiol.* **69** *(6) 3152-3157*

Tiedje, J.M. (1988). Ecology of denitrification and dissimilatory nitrate reduction to ammonia. In: Biology of anaerobic microorganisms. Zehnder, A.J.B (Ed)

UNEP, (1999). Planning and Management of Lakes and Reservoirs': An integrated Approach to Eutrophication. IETC Technical publication Series No 11. International Environmental Technology Centre, United Nations Environment Program, Japan United Nations (2004). World urbanization prospects: The 2003 Revision Data Tables and Highlights. New York

Van der Steen, N.P., Brenner, A., Shabtai, Y. and Oron, G. (2000a). Effect of environmental conditions on faecal coliform decay in post–treatment of UASB reactor effluent. *Wat. Sci. Tech.* **42** *(10-11); 111-118*

Van der Steen, N.P., Brenner, A., Shabtai, Y. and Oron, G. (2000b). Improved fecal coliform decay in integrated duckweed and algal ponds. *Wat. Sci. Tech.* **42** *(10-11); 363-370*

Veenstra, S. and Alaerts, G. (1996). Technology selection for pollution control. In: A. Balkema, H. Aalbers and E. Heijndermans (Eds.), Workshop on sustainable municipal waste water treatment systems, Leusdan, the Netherlands; 17-40

Yu, H., Tay, J. and Wilson, F. (1997). A sustainable municipal wastewater treatment process for tropical and subtropical regions in developing countries. *Wat. Sci. Tech.* **35** *(9); 191-198*

Zhao, Q., and Wang, B. (1996). Evaluation on a pilot-scale attached-growth pond system treating domestic wastewater. *Water Resource* **30**; *242-245*

Zhu, A.N., Zhang, J.B., Zhao, B.Z., Cheng, Z.H. and Li, L.P. (2005). Water balance and nitrate leaching losses under intensive crop production with Ochric Aquic Cambosols in North China Plain. *Environmental International 31; 904-912*

Zimmo, O.R., Al sa'ed, R. and Gijzen, H.J. (2000). Comparison between algae based and duckweed based wastewater treatment. Differences in environmental conditions and nitrogen transformations. *Wat. Sci. Tech. 42 (10-11); 215-222*

Zimmo, O.R., Steen, N.P. and Gijzen H.J. (2003). Comparison of ammonia volatilization rates in algae and duckweed based wastewater stabilization ponds. *Env. Tech. 25; 273-282*

Chapter 2
The effect of baffles on algal-bacterial biofilm structure and composition of zooplankton in wastewater stabilization ponds

Chapter 2

The effect of baffles on algal-bacterial biofilm structure and composition of zooplankton in wastewater stabilization ponds

Abstract

Nitrifier biomass in the water column of wastewater stabilization ponds is low and this reduces nitrification rates. For this study, baffles were installed as attachment surface for nitrifiers. Four pilot scale wastewater stabilization ponds were constructed in Kampala, Uganda; pond 1 was control while ponds 2, 3 and 4 had baffles of the same total surface area but different configurations. The ponds were operated for two periods i.e. under an influent BOD of 72 ± 45 mg l^{-1} and ammonia of 34 ± 7 mg l^{-1} (period 1) and an influent BOD of 29 ± 9 and ammonia of 51 ± 4 mg l^{-1} (period 2, achieved by covering facultative pond). The objective of this study was to investigate biofilm biomass and, distribution and composition of algae on the baffles. The diversity and biomass of zooplankton in the pond water column was also investigated. Dry weight of biofilm was determined gravimetrically while the wet weight of the algae in the biofilm were determined using the bio volume method. The results showed that the algal-bacterial biofilms grew rapidly within 2-3 weeks and after that, the thickness did not increase. The algal-bacterial biofilm dry weights at 5cm depth of all ponds during period 1 were higher than those of period 2. This is probably due to the contribution of influent algae and heterotrophic biomass. As the facultative pond during period 1 was not covered, more algae entered the maturation ponds forming thick biofilms on the baffles, the algae in the deeper layers died forming a layer of detritus material. Additionally, the higher influent BOD during period 1 might have favored growth of more heterotrophic bacteria. The dry weight measurements included all components of the biofilm such as algae, heterotrophic organisms, midge larvae and detritus while the wet weight specific to only algae, was measured by the bio volume method. This involved identifying algae under the microscope and calculating their wet weight using geometric formula. It was found that the total wet weight of the algal biofilm during period 2 were higher than in period 1, which was opposite to the dry weight results. Probably the dry weight during period 1 consisted to a large extent of dead algae material, which is excluded from the wet algae weight. The advantage of bio volume method is that wet weight of only living algae is determined. The reason why the total wet weights of biofilm algae were higher during period 2 can be attributed to covering of the facultative pond. During period 2, no algae entered the maturation ponds therefore the TSS in the water column was significantly lower than in period 1. This permitted more light penetration into the deeper parts of the ponds thus increasing the area for attached algal growth. Additionally, the influent ammonia during period 2 was higher; this coupled with no algae entering from the covered facultative pond could have reduced competition. More algal biomass during period 2 led to improved oxygen conditions in the ponds. The DO in all ponds at 45 and 70cm depth were significantly higher during period 2; this is important for ammonia oxidation. Algal species were identified and generally, there was a shift in the dominant group from period 1 to period 2 indicating a change in pond behavior. The zooplanktons were also diverse with young stages of copepod (Nauplius larvae) being dominant in all the four ponds. The distribution of algae and zooplankton in the four ponds showed that the baffles had an effect on water quality which in turn affected the ecology of wastewater stabilization ponds.

Key words: Stabilization pond, biofilm, baffle, algae, zooplankton

Introduction

Wastewater stabilization ponds (WSP's) are wastewater treatment systems that use mutual relationship between bacteria and algae (Kayombo *et al.,* 2002). These are the common treatment technologies used in most developing countries. This is due to their simplicity, low costs of construction, maintenance and operation (Veenstra and Alaerts, 1996). However, they are not effective in nutrient removal especially nitrogen and this has been attributed to low nitrifier biomass in the water column (McLean *et al.*, 2000; Zimmo *et al.*, 2000). To improve nitrification in these ponds, attachment surfaces for the nitrifiers have been used (Baskaran *et al.,* 1992; Zhao and Wang, 1996; McLean *et al.*, 2000; Schumacher and Sekoulov, 2002).

This study incorporated baffles as attachment surface in pilot scale wastewater stabilization ponds. The baffles were flat wooden plates coated with fibre glass material; they were arranged in different configurations in the different ponds. Introduction of baffles affects pond hydraulics and may change the pond ecology. WSP's are complex ecosystems whose performance is related not only to hydraulic characteristics, pollution loading and climate but also to biological communities like planktons and benthos. Physico-chemical variables alone cannot explain performance of WSP's. Approaches that take into account physico-chemical and biological variables and their interactions should be adopted (Cauche *et al.,* 2000). Wastewater stabilization ponds mostly depend on oxygen production by algae (Pearson, 2005) thus intensive algal grazing by zooplanktons can be detrimental to treatment process (Lai and Lam, 1997; Pearson, 2005). For eutrophic lakes, zooplanktons have been reported to release ammonia through excretion (Smith, 1978; Lehman, 1980; Moss, 1988, Moegenburg and Vanni, 1991) though this is rapidly taken up by algae (Ganf and Blazka, 1974; Moegenburg and Vanni, 1991). However, it is uncertain whether this has an effect on wastewater treatment in wastewater stabilization ponds.

Development of algal-bacteria biofilms are influenced by many factors like the nature of attachment surface, nutrients, light and other biological interactions. Different types of algae prefer different growth conditions; their dominance and distribution on the biofilm differs accordingly. For instance, Cyanobacteria and Euglena are tolerant to high organic and nutrient loading (Wrigley and Torien, 1990; Sheheta and Badr, 1996). The motile forms like Euglena and Chlamydomonas are known to migrate to areas with sufficient illumination (Moss, 1988). The phytoplankton distribution and occurrence can also be determined by grazing pressure of the herbivorous zooplanktons (Michael, 1987; Bakker and Rijswijk, 1994; Lai and Lam, 1997).

Zooplanktons tend to occur in ponds with low BOD loading (Uhlman, 1980). There are three major crustacean zooplanktons of ponds viz. copepods, cladocera and rotifers. These have different sizes and different feeding habits although overlap in some instances. The adult copepods are larger than cladocerans while the rotifers are the smallest in the group. The copepods can be small-particle feeders (cyclopoid copepods) but generally feed on a wider range of food particles (5-100μm). Copepods can prey on smaller zooplanktons, larger colonies or masses of phytoplankton. They have eleven successive moults before maturation into adult copepods; the first six moults are juveniles' referred to as nauplii (Moss, 1988).

The cladocera have a carapace which covers their bodies; they can be filter feeders or raptorial i.e. actively grasp their prey e.g. *Leptodora* and *Polyphemus*. The herbivorous genera include *Daphnia* and *Bosmina*. The filter feeders take up food particles in the range of 1-50 μm. The particle feeders have thoracic limbs with hairs which convey food to their mouth (Moss, 1988). The cladocera actively move through water using their large branched second antennae giving them the common name of water fleas. Rotifers are generally smaller in size and usually feed on small particles 1-20 μm in size. They are suspension feeders; they have rhythmically beating cilia around the mouth which directs water and suspended particles into their gut. Knowledge of zooplanktons of wastewater stabilization ponds is limited as compared to information available on microalgae. The zooplankton and insect larvae are known to play important role in the treatment process of activated sludge and biological filters. Daphnia can polish algal rich effluents and could have applications in aquaculture (Pearson, 2005).

The objective of this study was to investigate biofilm development on the baffles installed in four pilot scale maturation ponds. Biofilm biomass, distribution and composition of algae on the baffles were investigated. The diversity and biomass of zooplankton in the pond water column of the pilot scale maturation ponds was further investigated.

Methodology
Description of pilot scale system
The pilot scale plant was set up at the Bugolobi Sewage Treatment Works (BSTW) in Kampala- Uganda. The plant consisted of three systems: (i) an anaerobic tank (AT) of 10 m³ which fed the (ii) facultative pond (FP) (10m x 2.0m x 1.0m) at a flow rate of 2.1 m³ per day and (iii) 4 fiber glass maturation ponds (MP) with different surface area for bacterial attachment and different flow patterns. The total surface area for attachment in pond 1 (un-baffled) was 6.4 m² while that of ponds 2, 3 and 4 was 23.2 m² each. The surface area for attachment in pond 1 was provided by the walls of the ponds while the extra surface area in ponds 2, 3 and 4 was provided by installation of different configurations of 15 baffles of the same area (0.56 m² each).

Each maturation pond had dimensions of 4.0 m x 1.0 m x 1. 0 m (length, width and depth) but wastewater was filled up to a depth of 0.8m. Pond 1 was a long open channel with horizontal flow. Pond 2 had fifteen baffles placed vertically and parallel to the flow direction and has horizontal flow pattern, similar to that of pond 1. Pond 3 also had fifteen four ponds were sampled during day time using a one liter plastic container. Each sample was filtered through a 53 μm sieve, stored in different sample bottles and preserved with 4% formaldehyde. The samples were transferred to the laboratory for analysis as described by Ndawula et al., (2004). In the laboratory, the samples were washed over a 53 μm sieve to remove the fixative, re-diluted into sufficient volume to obtain workable counting densities. Sub-samples of 2ml, 5mls and 10mls were taken from well agitated samples to ensure homogeneous distribution of organisms. The sub-samples were introduced on a counting chamber and examined under an inverted microscope at both X100 magnification for taxonomic analysis and X40 magnification for counts to determine species composition and abundance, respectively. Identifications were done to the lowest possible taxon using published keys and figures (Sars, 1895; Rzoska, 1957; Brooks, 1957; Ruttner, 1974; Pennak, 1978). Dry-biomass and densities were generated using the bio volume as described in Ndawula (1998) and Manca and Comoli, (2000).

Results

The results for algal-bacterial biofilm dry weight at 5cm depth after 5 weeks are shown in figure 2.1. The dry weight biomass at only 5cm depth can be compared between the two periods (those of 45 and 70cm were not determined during period 1, see methodology). The results for wet weight biofilm algal biomass are presented in table 2.1 and these can be used for comparison at the three depths of the two periods.

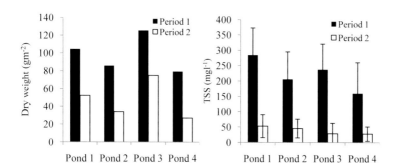

Figure 2.1: Algal-bacterial biofilm dry weight at 5cm depth after 5 weeks incubation for periods 1 and 2 in maturation ponds 1-4

Figure 2.2: TSS for periods 1 and 2 in maturation ponds 1-4 at 5 cm depth (n = 61, 54)

Table 2.1 Wet weight biofilm algal biomass (gm^{-2}) for the two periods estimated using bio volume of algal cells

Pond	Period 1 (gm^{-2})				Period 2 (gm^{-2})			
	5cm	45cm	70cm	Total	5cm	45cm	70cm	Total
Pond 1	1.03	0.65	0.49	2.2	0.73	1.06	0.16	2.0
Pond 2	5.74	1.66	0.93	8.3	7.13	10.4	0.51	18.0
Pond 3	2.93	4.99	0.13	8.1	7.75	3.79	1.13	12.7
Pond 4	0.16	0.06	1.22	1.4	8.57	9.99	6.06	24.6

TSS

The results for the effluent TSS of the maturation ponds are presented in figure 2.2. The effluent TSS of ponds 1, 2, 3 and 4 during period 1 were 285± 89, 206 ± 90, 238 ± 83 and 160 ± 100 mg l^{-1} respectively. Kruskal-Wallis test showed a significant difference between the effluent TSS of pond 1 and 2 ($p < 0.0001$); but TSS of pond 3 did not significantly differ from these two. The TSS of pond 4 was significantly lower than that of ponds 1 and 3 ($p < 0.0001$). The effluent TSS of pond 1, 2, 3 and 4 during period 2 were 53±37, 46±30, 29±34 and 28±23 mg l^{-1} respectively. Non parametric test showed that the median TSS of pond 1 and 2 were not significantly different, similar results were obtained for ponds 3 and 4. However, the former were significantly different from the latter. Comparisons between the two periods showed that the TSS during period 1 was significantly higher than those of period 2. The higher TSS during period 1 could have been partly due to entry of suspended algae from the facultative pond (during period 1, the surface of the facultative pond was not covered so the influent to the maturation pond included algae).

Dry Weight Biomass

During period 1, pond 3 had the highest algal-bacterial biofilm dry weight followed by ponds 1, 2 and 4; the same order was observed during period 2 (Figure 2.1). Although the biofilm biomass of pond 3 appeared to be higher than the other ponds in both periods, it was difficult to make conclusions if it was significantly high due variations in data. In terms of biofilm growth rates, pond 3 had the highest algal-bacterial biofilm growth rates in both period 1 and 2 i.e. 3.1 and 3.6 gm^{-2} d^{-1} respectively. When each pond was compared during the two periods, the algal-bacterial biofilm dry weights of period 1 were generally higher than those in period 2 (Figure 2.1).

The results for algal-bacterial biofilm biomass (dry weight) development with time for period 2 at three depths of 5cm, 45cm and 70 cm are shown by figures 2.3-2.6. It was also seen that the biomass in pond 1, 2 and 3 were very dynamic; biofilm formation and sloughing occurred simultaneously. Although the biofilm biomass at 5cm depth was most of the time the highest, t-tests showed no significant difference between the biomass of this depth and that of 45 and 70cm in all ponds except for pond 1 at 5 and 45cm depth. At 5cm depth, pond 3 had the highest biomass of 76 g m^{-2} (at week 3), for the other ponds the highest values are: 53 g m^{-2} (pond 1, week 5), 36 g m^{-2} (pond 2, week 3) and lastly 27g m^{-2} (pond 4, week 5). The results showed that the ponds behaved differently although they received the same influent. Generally, the biomass in pond 4 at all depths was lower than those of ponds 1, 2 and 3.

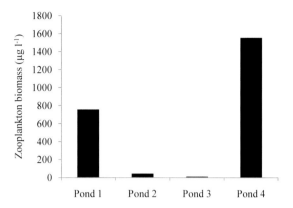

Figure 2.9: Total zooplankton biomass on basis of size of organism in the four maturation ponds during period 2

Discussion
Dry weight biomass
The results for both period 1and 2 showed that the biofilm biomass dry weight in pond 3 was higher than in the rest of the ponds. However, it was difficult to get clear differences between the dry weights of the four ponds. This is due to the variation of dry weights which could have been caused by different components of the biofilm i.e. algae, heterotrophs, midge larvae and detritus. The different components and their relative contribution to the dry weight may have varied in the different ponds hence causing the differences. However, comparison of dry weights between the two periods showed that algal-bacterial biofilm dry weight of period 1 were higher than those of period 2 (Figure 2.1). Visual inspection of the biofilm during algal counts and identification showed that the biofilm during period 1 (as compared to that of period 2) was thicker in size and mainly consisted of detritus material. This could be due to importation of algae from the facultative pond. During period 1, the surface of the facultative pond was open hence more algae entered the maturation pond; it is possible that more algae attached to the biofilm plates forming thick layers. When biofilms develop and become thick, the inner layers become inactive and die due to light and nutrient limitation (Lazarova and Manem, 1995). The inactive layers become detritus material which supplies more organic matter to heterotrophs via biomass lyses (Wolf *et al.,* 2007). Barranguet *et al.,* (2005) reported that when algal-bacterial biofilm develop under low light, the proportion of heterotrophs to algae increases. The bulk water TSS of period 1 was significantly higher than period 2 (Figure 2.2) and this could have reduced light penetration during period 1. Under these conditions, there is a possibility that the heterotrophic biomass in period 1 was higher than that of period 2. The other explanation for the differences in algal-bacterial biofilm dry weights during the two periods could be due to the differences in influent BOD concentrations which resulted from change of operational conditions. It was found that filtered influent BOD of the maturation ponds during period 1 (72 ± 45.1 mg l^{-1}) was significantly higher (p<0.0001) than that of period 2

(29.4±9.2 mg l^{-1}). The higher influent BOD during period 1 is thought to have favored growth of more heterotrophic bacteria partly accounting for the higher dry weight during this period.

The results for biofilm biomass of the maturation ponds during period 2 depicted fluctuations in the biomass with time as can be seen in figures 2.3 – 2.6. The reasons for this are not clear; it is thought to be due to changing pond conditions or to the effect of grazing by zooplankton. Midge larvae of Chironomidae made their cases in the algal biofilm (Figure 2.8) probably dislodging and feeding on biofilm matter (Lamberti, 1996). The results showed the dynamic nature of the algal-bacterial biofilms of WSP; probably steady state biofilms in these systems are short lived or are not reached at all. This may be of advantage to wastewater treatment i.e. in terms of renewal of biofilms or to the disadvantage i.e. dislodging of the already established attached biomass. The effect of this phenomenon requires further research.

Wet weight biomass
Two key observations were made from the results of the wet weight algal biofilm: First, all the values of wet weight of the algal biofilm (Table 2.1) were lower than those of dry weight algal-bacterial biofilm (Figure 2.1). Secondly, the algal wet weight biofilm during period 2 were higher than in period 1 (Table 2.1); results which are opposite to those of dry weight algal-bacterial biofilm. The wet weight algal biomass is specific to only algae while the algal-bacterial biofilm dry weight consists of dry weight of algae, heterotrophic organisms, and midge larvae and detritus material. This led to higher algal-bacterial dry weight values. Since the wet weight of the algal biofilm represents the living algae in the biofilm, the percentage of active algae at 5cm depth present in the dry weight can be estimated. The percentage of living algae present in the dry weight during period 1 were 1%, 7%, 2% and 0.2% in ponds 1, 2, 3 and 4. Those of period 2 were 1.4%, 21%, 10% and 32% for ponds 1, 2, 3 and 4 respectively. The percentage weight of algal material would be accurately estimated by chlorophyll a measurement; this is recommended for future studies. The advantage of using the bio volume method for biomass estimation was that information on both biomass and algal types were obtained concurrently.

The total wet weight of algal biofilm of pond 1 during period 1 and 2 did not differ while those of ponds 2, 3 and 4 were higher during period 2 (Table 2.1). It was seen that the algal biomass in the deeper parts of the biofilm substantially increased; especially in pond 4. This may be attributed to the significantly lower bulk water TSS of period 2 (Figure 2.2) which could have allowed more light penetration in the deeper parts of the ponds. Additionally, period 2 had higher influent ammonia therefore more possibilities for algal growth and higher biofilm biomass. However, increase in influent ammonia would equally affect both biofilm and bulk water algae but the bulk water TSS of the maturation ponds in period 2 was lower than in period 1. The lower bulk water TSS during period 2 could have been caused either by absence of algae in the influent or due to the grazing pressure of zooplanktons on algae. The surface of the facultative pond was covered during period 2 and this did not permit algae to develop. Zooplanktons are known to prevail under more aerobic conditions (Pearson, 2005) which existed during period 2. Increase in oxygen suggests more algae and under such conditions more zooplankton, which could have kept the biomass under control (algae and heterotrophs), so it seems there was a self controlling system.

Effect of algal biomass on oxygen production

The oxygen concentrations of the ponds during period 2 significantly improved especially at 45cm and 70cm pond depths (Figure 2.7). This is consistent with the wet biofilm algal biomass which was higher in the deeper parts of the ponds during period 2. The improvement of oxygen conditions and lower BOD during period 2 is important for nitrogen removal in WSP. These conditions could favor growth of nitrifiers and this may have had a positive effect on ammonia oxidation. The oxygen concentration in ponds 1, 2 and 3 at 45cm depth during period 2 were above 5 mg l^{-1}. These results are in agreement Kayombo *et al.,* (2002) who found daily average oxygen levels at 30cm depth in maturation ponds to range from 3.4 to 11.3 mg l^{-1}. The maturation ponds have been shown to be aerobic during day time and anaerobic during night time (Kayombo *et al.,* 2002). This implies that ammonia oxidation occurs during the aerobic phase followed by denitrification during the anaerobic phase. The depth at which aerobic conditions exist in ponds is important with respect to nitrogen removal; this increases both the aerobic volume and area (on biofilms) for nitrifiers. To understand nitrogen removal in these ponds, it is vital to study nitrification rates at different oxygen conditions and this can be related to oxygen conditions in ponds.

Algal composition in biofilms

The dominant algae groups (based on wet biomass) in the biofilm during period 1 in pond 1 at 5cm depth were Chlorophyta (*Chlamydomonas* sp), pond 2 and 3 Cyanobacteria (*Planktolyngbya* sp) and pond 4 Euglenophyta (*Euglena* sp). *Chlamydomonas s*p and *Euglena* sp are motile algae and are least expected to be found attached on biofilms. Probably they attached to prevent from being washed out of the ponds. The motile forms especially *Chlamydomonas* sp is known to dominate ponds with turbid water (Mara and Pearson, 1987) and can tolerate high organic loading. *Chlamydomonas* sp exhibit high tolerance to sulphide which cannot be said of *Euglena* sp (Pearson, 2005). The results from this study showed that the bulk water TSS in pond 1 (which was significantly higher than in the other ponds) caused turbidity which favored *Chlamydomonas* sp. The dominant groups in ponds 1 and 2 at 45cm were Cyanobacteria (*Planktolyngbya* sp), pond 3 Chlorophyta (*Gongosira* sp) and pond 4 Diatoms (*Gomphocybella* sp). At deeper depths of 70cm, Euglenophyta was dominant in pond 1 while Cyanobacteria (*Planktolyngbya* sp) dominated in ponds 2, 3 and 4. It is known that Cyanobacteria and Euglenophytes thrive under high organic and nutrient rich water (Moss, 1988; Sheheta and Badr, 1996). In terms of algal species found in the biofilms, the combined numbers of species (5cm, 45cm and 70cm) were 11, 15, 8 and 21 algal species in ponds 1, 2, 3 and 4 respectively (Table A1, Appendix). The diversity of algae is known to increase with low organic loading; pond 4 had the lowest effluent BOD.

The dominant algae on the biofilm at 5cm, 45cm and 70 cm depth in pond 1 during period 2 were Cyanobacteria (*Planktolyngbya* sp) and Chlorophyta (*Protoderma* sp), table 2.2.

Those of ponds 2 were Euglenophyta (*Trachelomonas* sp and *Phacus longicuada*), pond 3 Cyanobacteria (*Planktolyngbya* sp *and Planktolyngbya limnetica*) and Euglenophyta and pond 4 Cyanobacteria (*Planktolyngbya* sp, *Plankotothrix* sp *and Planktolyngbya limnetica*). It was found that Cyanobacteria were most common at 5cm depth of the ponds and this could be due to selective grazing pressure by zooplanktons. The zooplanktons tend to feed on the smaller algae and avoid the toxic Cyanobacteria (Gilbert, 1996). The total number of algal species during period 2 in ponds 1, 2, 3 and 4 were 17, 21, 17 and 14 (Table A2, appendix). The species diversity was higher during period 2 (except pond 4) implying that the water quality had improved in terms of organic loading (Pearson, 2005). Generally, there was a shift in the dominant group of algae from period 1 to period 2 indicating that the pond behavior was changing.

Zooplankton composition
Small bodied cladocerans like the *Ceriodaphnia* and *Miona* are known to be insensitive to Cyanobacteria toxins (Hanazato, 1991; Bouvy *et al*., 2001). It is possible that they were able to withstand the Cyanobacteria that dominated pond 1 and 4 during period 2 (Table 2.2). *Miona* also thrives under favorable trophic conditions created by fragmentation of Cyanobacterial filaments by copepods and rotifers (Bouvy *et al*., 2001). Other, large cladoceran species like *Daphnia* and *Diaphanosoma* are more sensitive to toxic filaments of Cyanobacteria (Kirk and Gilbert, 1992) and this explains their absence in the ponds. It was seen that Cyanobacteria also dominated pond 3 but *Ceriodaphnia* and *Miona* were not present. This can be attributed to predation pressure; it is known that the last stages of copepods and adults cyclopoids are carnivorous (Brandl, 1998; Bouvy *et al.,* 2001). The higher numbers of Nauplius larvae in pond 2 and 3 may have suppressed the population of *Miona micrura* in these ponds.

There are a number of reasons for existence of rotifer *Branchionus angularis* in only ponds 1 and 2. These include tolerance to pollution and toxins, food availability and competition, predation and mechanical disturbance. For instance *Branchionus angularis* has been found to be most tolerant to pollution (Sladecek, 1983) and toxins of Cyanobacteria (Fulton and Paerl, 1988; Kirk and Gilbert, 1992). Pond 1 had the highest effluent ammonia of 27.8±4.4 mg l^{-1}. Rotifers are small organisms which prefer to feed on smaller food particles like bacteria, detritus and algae (Starkweather, 1980). Smaller and palatable algal forms like Chlorophyta and Euglenophytes were among the dominant groups during period 2 in ponds 1 and 2 (Table 2.2) and these favored *Branchionus* rotifers in these ponds. It is also known that copepods and rotifers compete for the palatable forms of algae hence presence of more copepod nauplii in pond 3 disadvantaged the rotifers.

It is reported that copepods prey on rotifers and further still, the larger zooplanktons mechanically damage the rotifers (Gilbert and Stemberger, 1985; Burns and Gilbert, 1986, Gilbert, 1988). Possibly the lower number of nauplii in pond 4 favored the existence of *Branchionus plicatilis* in this pond. *B.plicatilis* can tolerate low oxygen concentrations (Barrabin, 2000); the oxygen concentrations of pond 4 at both 45 and 70cm were significantly lower than the other ponds and this could have favored them. Most of the zooplanktons recovered in the samples were grazers of algal cells especially *Brachionus*,

Moina and *Cyclops*. Studies by Michael (1987), Moegenburg and Vanni, (1991); Bakker and Rijswiik, (1994) and Lai and Lam, (1997) show that zooplankton can exert pressure on algal populations through grazing. Possibly the smaller algal forms were grazed upon (Bouvy *et al.,* 2001) and this is one of the reasons for Cyanobacteria being common in the ponds.

The effect of zooplankton in wastewater treatment has been largely discounted (Mitchell and Williams, 1982) yet they have been found to release nutrients back into water (Smith, 1978; Lehman, 1980; Lampert *et al.*, 1986; Moss, 1988; Lai and Lam, 1997). For instance, Smith, (1978) found ammonia released by fed copepods to be 0.303 µg N individual^{-1} day^{-1}. The smaller herbivores like rotifers and protozoan have high metabolic rates and can release even more nitrogen; for instance they have been found to release up >35-60% of their own phosphorous content (Lehman, 1980). To demonstrate the effect of zooplankton on wastewater treatment, the average number of nauplius larvae per liter of wastewater for the four ponds can be calculated from table 2.3. These can be used to calculate the total number of individuals for the whole pond volume of 3200 liters and using the ammonia release rate of Smith, (1978); the ammonia release per day by the zooplankton can be calculated. For instance, 10560, 26560, 35200 and 4800 nauplii were calculated for ponds 1, 2, 3 and 4 respectively. These gave the amount of ammonia released as 3.2, 8.0, 10.1 and 1.4 mg-N d$^{-1,}$ which is substantially lower that the influent ammonia load of 26,520 mg N d^{-1}. Probably if the ammonia released by all the other zooplankton is considered, the amount of ammonia released may be important. The algal uptake of ammonia excreted by zooplankton and subsequent feeding of algae by zooplankton shows internal recycling of nitrogen. As long as the zooplanktons remain in the ponds, they don't contribute in the nitrogen removal. Generally, the effect of zooplankton on nitrogen cycle in wastewater stabilization ponds is not fully understood and further studies in this area is recommended.

Conclusions
Generally, the algal bacterial biofilm dry weight biomass during period 1 was higher than those of period 2. The difference could have been caused by more algae entering from the facultative pond and attaching on baffles to form thick layers. Death of the algae in the deeper layers could have increased the proportion on detritus material during period 1. The conditions during period 1 i.e. less light, more influent BOD and detritus material could have favored more heterotrophs during this time. Also the oxygen levels during period 2 were higher and this could have favored the zooplankton which grazed on the biofilm. During period 2, the biofilm biomass varied with time, a constant phase was not attained as expected. The biomass rose and dropped, probably caused by changes in pond environment, zooplankton grazing or disturbance by the blood worms. The studies showed that biofilm development can be attained in three weeks and it is possible to use biofilms in wastewater stabilization ponds.

The wet algal biomass during period 2 was higher than in period 1; this caused significantly higher oxygen concentration in the deeper parts of the ponds during period 2. This could have created favorable conditions for growth of nitrifiers since competition from heterotrophs is minimized by low BOD loading during period 2.

The algal diversity during period 1 was lower than that of period 2; an indication that the water quality of the ponds improved during this time. The covering of FP during period 2 led to low BOD loading which improved the water quality. Low BOD is also known to favor growth of nitrifiers which lead to more ammonia removal.

The studies also showed that there was diversity of zooplankton in the ponds which may have been as a result of algal distribution, physico-chemical or hydraulic characteristics of the ponds. However, the role of zooplankton in the nitrogen cycle in wastewater stabilization ponds is still unclear; further investigations are recommended.

Acknowledgements
We would wish to extend our sincere thanks to the Netherlands Government (through NUFFIC) and EU- SWITCH project number 018530 for providing financial assistance. We also thank Aguzu Alex and Kigundu Vincent from the National Fisheries Research Institute (NaFFIRI), Jinja – Uganda for the assistance in identifying phytoplankton and zooplanktons. Their contributions to this work are invaluable.

References
APHA. (1995). Standard Methods for Examination of Water and Wastewater 19th Ed., Washington, D.C

Bakker, C. and Van Rijswijk, P. (1994). Zooplankton biomass in Oosterschelde (SW Netherlands) before, during and after the construction of a storm-surge barrier. *Hydrobiologia, 282(283), 127-143*

Barrabin, J.M. (2000). The rotifers of Spanish reservoirs: Ecological, systematical and zoogeographical remarks. *Limnetica 19, 91-167*

Barranguet, C., Veuger, B., Beusekom, S.A.M., Marvan, P., Sinke, J.J., Admiral, W. (2005). Divergent composition of algal-bacterial biofilms developing under various external factors. *Eur.J.Phycol. 40, 1-8*

Baskaran, K., Scott, P.H. and Connor, M.A. (1992). Biofilms as an Aid to Nitrogen Removal in Sewage Treatment Lagoons. *Wat. Sci. Tech. 26 (7-8), 1707-1716*

Brandl, Z. (1998). Feeding strategies of planktonic cyclopoids in lacustrine ecosystems. *J. Mar. Syst 15, 87-95*

Brooks, J. l. (1957). The systematic of North American Daphnia. *Mem. Conn Acad Arts Sci*

Bouvy, M., Pagano, M. and Trousellier, M. (2001). Effects of a cyanobacterial bloom (*Cyclindrospermopsis raciborskii*) on bacteria and zooplankton communities in Ingazeira reservoir (northeast Brazil). *Aquat Microb Ecol. 25, 215-227*

Burns, C.W and Gilbert, J.J. (1986). Effects of daphnid size and density on interference between *Daphnia* and *Keratella cochlearis*. *Limn. Ocean.* *31, 848-858*

Cauchie, H.M.; Salvia, M; Jean-Pierre Thome and Hoffmann, L. (2000). Performance of a single-cell aerated wastewater stabilization pond treating domestic wastewater: A Three year Study. *Internat. Rev. Hydrobiol* *85, (2-3)231-251*

Fulton, R.S and Paerl, H.W. (1988). Effects of the blue-green algae on *Microcystis aeruginosa* on zooplankton competitive relations. *Oecologia* *76, 383-389*

Ganf, G.G. and Blazka, P. (1974). Oxygen uptake, ammonia and phosphate excretion by zooplankton of shallow equatorial lake (Lake George, Uganda). *Limn. Ocean.* *19 (2), 313-325*

Gilbert, J.J. and Stemberger, R.S. (1985). Control of *Keratella* populations by interference competition from *Daphnia*. *Limn. Ocean.* *15, 839-928*

Gilbert, J.J. (1988). Susceptibilities of ten rotifers species to interference from *Daphnia duplex*. *Ecology* *69, 1826-1838*

Gilbert, J.J. (1996). Effect of food availability on the response of planktonic rotifers to a toxic strain of the cyanobacterium *Anabaena flos-aquae*. *Limn. Ocean.* *42, 1565-1572*

Hanazato, T (1991). Interrelations between Microcystis and Cladocera in highly eutrophic Lake Kasumigaura, Japan. *Int Rev Hydrobiol.* *76, 21-36*

Hillebrand, H., Durselen, C.D., Kirschtel, D., Pollingher, U., Zohary, T. (1999). Biovolume calculation for pelagic and benthic microalgae. *Journal of Phycology* *35: 403–424*

Kayombo, S., Mbwette, T.S.A., Mayo, A.W., Katima, J.H.Y., Jørgensen, S.E., (2002). Diurnal cycles of variation of physical–chemical parameters in waste stabilization ponds. *Ecol. Eng (18), 287-291*

Kirk, K.L. and Gilbert, J.J (1992). Variation in herbivore response to chemical defenses: zooplankton foraging on toxic Cyanobacteria. *Ecology* *73, 2208-2217*

Lai, P.C.C and Lam, P.K.S. (1997). Major Pathways for Nitrogen Removal in Wastewater Stabilization Ponds. *Water, Air and Soil pollution,* *94, 125-136*

Lamberti, G. A. (1996). The role of periphyton in benthic food webs. In Stevenson, R. J., M. L. Bothwell & R. L. Lowe (Eds), Algal Ecology. Academic Press, San Diego: 533–572

Lampert, W., Fleckner, W., Rai, H. and Taylor, B.E. (1986). Phytoplankton control by grazing zooplankton: A study on the spring clear-water phase. *Limnol. Ocean.* *31 (3), 478-490*

Lazarova, V and Manem, J. (1995). Biofilm characterization and activity analysis in water and wastewater treatment. *Wat. Res.* **29** *(10), 2227-2245*

Lehman, J.T. (1980). Release and cycling of nutrients between planktonic and herbivores. *Limn. Ocean.* **25(4)***, 75-83*

Manca, M. and Comoli, P. (2000). Biomass estimates of freshwater zooplankton from length-carbon regression equations. *Journal of Limnology,* **59***(1), 15-18*

Mara, D.D. and Pearson, H.W. (1987). Waste stabilization ponds. Design Manual for Mediterranean Europe

McLean, B.M., Baskran, K. and Connor, M.A. (2000). The use of algal-bacterial biofilms to enhance nitrification rates in lagoons: Experience under laboratory and pilot scale conditions. *Wat. Sci. Tech.* **42** *(10-11); 187-194*

Michael, J.V. (1987). Effects of Nutrients and Zooplankton size structure of a phytoplankton community. *Ecology:* **68** *(.3), 624 – 635*

Mitchell, B.D. and Williams, W.D. (1982). Population Dynamics and Production of *Daphnia carinata* (King) and *Simocephalus exspinosus* (Koch) in Waste Stabilization ponds. *Aust. J. Mar. Freshw. Res,* **33***, 837-864*

Moegenburg, S.M and Vanni, M.J. (1991). Nutrient regeneration by zooplankton: effects on nutrient limitation of phytoplankton in a eutrophic lake. *Journal of Plankton Research,* *31 (3), 573-588.*

Moss, B. (1988). Ecology of Fresh Waters; Man and Medium 2nd edition, Blackwell Scientific Publication

Ndawula, M.L., Kiggundu, V. and Ghandhi, P.W. (2004). The status and significance of invertebrate communities. In: Challenges for Management of the Fisheries Resources, Biodiversity and Environment of Lake Victoria. Editors: J. S. Balirwa, R. Mugidde, R. Ogutu-Ohwayo. Fisheries Resources Research Institute, Technical Document No. 2, First edition, 153 – 171

Ndawula, M.L. (1998). Distribution, abundance of zooplankton and *Rastrineobola argentea* (Pisces: *Cyprinidae*) and their trophic interactions in northern Lake Victoria, East Africa. PhD. Thesis, University of Vienna, Austria

Pagano, M. (2008). Feeding of Tropical Cladocerans (*Moina micrura, Diaphanosama excisum*) and Rotifer (*Branchionus calyciflorus*) on natural phytoplankton: Effect of phytoplankton size-structure. *J. Plank Res.* **30** *(4), 401-414*

Pearson, H.W. (2005). Microbiology of waste stabilization ponds. In: Pond Treatment Technology (Ed). Shilton, IWA publishing

Pennack, R.W. (1978). Fresh-water invertebrates of the United States. 2nd Edition, Wiley-Inter science. New York, NY

Rutner, K.A. (1974). Planktonic rotifers: Biology and taxonomy. Biological Station Lunz of the Austrian Academy of Science. E. Schweizerbart'sche Verlagsbuchhandlung, 129 pp

Rzoska J. (1957). Notes on the crustacean plankton of Lake Victoria. *Proc. Limn. Soc. Lond.* **168**, *1126-125*

Sars, G.O. (1895). An account of the Crustacea of Norway. Vol 1 Amphipoda. Description pp. 1-711. Christiania and Copenhagen. (Alb. Cammermeyer's Forlag).13: 1 - 1 8
Schumacher, G. and Sekoulov, I. (2002). Polishing of secondary effluent by an algal biofilm. *Wat. Sci. Tech.* **46** *(8); 83-90*

Shanthala, M., Shankar, P.H. and Hosetti, B.B. (2009). Diversity of phytoplankton's in a waste stabilization pond at Shimoga Town, Karnataka State, India. *Environ. Monit. Assess,* **151**, *437-443*

Sheheta, S.A. and Badr, S.A. (1996). Planktonic algal populations as an integral part of wastewater treatment. *Environmental management and Health 7/1 p 9-14*

Smith, L.S. (1978). The Role of Zooplanktons in the Nitrogen Dynamics of a Shallow Estuary. *Estuarine and Coastal Marine Science, 7, 555-565*

Sladecek, V. (1983). Rotifers as indicators of water quality. *Hydrobiologia* **100**, *169-201*

Starkwheather, P.L. (1980). Aspects of the feeding behavior and trophic ecology of suspension-feeding rotifers. *Hydrobiologia 73, 63-72*

Ulman, D. (1980). Limnology and performance of waste treatment lagoons. *Hydrobiologia* **72**, *21-30*

Veenstra, S. and Alaerts, G. (1996). Technology selection for pollution control. In: A. Balkema, H. Aalbers and E. Heijndermans (Eds.), Workshop on sustainable municipal waste water treatment systems, Leusdan, the Netherlands; 17-40

Wolf, G., Picioreanu, C., Loosdrecht, M.C.M. (2007). Kinetic modeling of phototrophic biofilms, The PHOBIA model. *Biotechnology and Bioengineering 97 (5), 1064-1079*

Wrigley, T.J. and Toerien, D.F. (1990). Limnological aspects of small sewage ponds. *Wat. Res.* **24** *(1), 83-90*

Zhao, Q., and Wang, B. (1996). Evaluation on a pilot-scale attached-growth pond system treating domestic wastewater. *Water Resource 30; 242-245*

Zimmo, O.R., Al sa'ed, R. and Gijzen, H.J. (2000). Comparison between algae based and duckweed based wastewater treatment. Differences in environmental conditions and nitrogen transformations. *Wat. Sci. Tech. 42 (10-11); 215-222*

Appendix 1: List 1 Algal composition on biofilm during period 1

Algal species	Biomass (gm^{-2})	Algal species	Biomass (gm^{-2})
Pond 1		**45cm depth**	
5cm depth		**Cyanobacteria**	
Cyanobacteria		*Merismopedia tenuissima*	9.8×10^{-3}
Merismopedia tenuissima	3.1×10^{-5}	*Planktolyngbya* sp	9.4×10^{-1}
Planktolyngbya contarta	1.1×10^{-4}	**Cryptophytes**	
Euglenophytes		*Cryptomonas* sp	5.6×10^{-2}
Euglena acus	4.6×10^{-2}	**Diatoms**	
Strombomonas acuminatus	1.8×10^{-2}	*Diatoma* sp	7.0×10^{-2}
Chlorophyta		*Fragilaria* sp	5.4×10^{-2}
Chlamydomonas sp	9.6×10^{-1}	*Gomphocybella* sp	1.3×10^{-1}
		Dianoflagelates	
45cm depth		*Glenodinium* sp	1.1×10^{-1}
Cyanobacteria		**Euglenophytes**	
Merismopedia tenuissima	7.0×10^{-4}	*Euglena* sp	1.2×10^{-1}
Planktolyngbya sp	4.9×10^{-1}	*Trachelomonas* sp	1.6×10^{-1}
Diatoms		**Chlorophyta**	
Gomphocybella sp	3.6×10^{-2}	*Scenedesmus* sp	1.4×10^{-2}
Nitzschia sp	9.3×10^{-4}	*Tetraedron trigonium*	3.4×10^{-3}
Euglenophytes			
Euglena sp	7.6×10^{-4}	**70cm depth**	
Chlorophyta		**Cyanobacteria**	
Protoderma sp	4.9×10^{-2}	*Planktolyngbya* sp	7.4×10^{-1}
		Cryptophytes	
70cm depth		*Cryptomonas* sp	8.5×10^{-2}
Cyanobacteria		**Diatoms**	
Merismopedia tenuissima	2.4×10^{-3}	*Diatoma* sp	2.2×10^{-2}
Planktolyngbya sp	2.3×10^{-1}	*Gomphocybella* sp	8.0×10^{-2}
Diatoms		*Nitszchia* sp	9.6×10^{-3}
Nitzschia sp	1.4×10^{-3}		
Euglenophytes		**Pond 3**	
Euglena sp	2.4×10^{-1}	**5cm depth**	
Trachelomonas sp	1.6×10^{-2}	**Cyanobacteria**	
		Planktolyngbya sp	2.6
Pond 2		*Planktothrix* sp	1.2×10^{-2}
5cm depth		**Diatoms**	
Cyanobacteria		*Diatoma* sp	1.3×10^{-2}
Planktolyngbya sp	4.8	*Gomphocybella* sp	1.7×10^{-2}
Planktothrix sp	9.9×10^{-2}	**Chlorophyta**	
Diatoms		*Gongosira* sp	2.2×10^{-1}
Gomphocybella sp	4.9×10^{-2}	*Protoderma* sp	3.2×10^{-2}
Nitzschia acicularis	1.1×10^{-2}	**45cm depth**	
Euglenophytes		**Cyanobacteria**	
Euglena sp	3.6×10^{-1}	*Planktolyngbya limnetica*	5.5×10^{-1}
Trachelomonas sp	4.3×10^{-2}	*Planktolyngbya* sp	1.8×10^{-3}
Chlorophyta		**Diatoms**	
Protoderma sp	3.0×10^{-1}	*Diatoma* sp	5.5×10^{-2}
Pseudodendoclonum sp	4.6×10^{-2}	*Gomphocybella* sp	1.8×10^{-1}
Tetraedron trigonium	6.2×10^{-3}	**Chlorophyta**	
		Gongosira sp	3.6

Continuation list 1, algal composition on biofilm during period 1

Algal species	Biomass (gm^{-2})	Algal species	Biomass (gm^{-2})
Protoderma sp	5.7x10^{-1}	**Diatoms**	
		Gomphocybella sp	4.3x10^{-1}
70cm depth		*Navicula gastrum*	8.0x10^{-2}
Cyanobacteria		*Nitzschia* sp	1.6x10^{-2}
Planktolyngbya sp	1.1x10^{-1}	**Euglenophytes**	
Diatoms		*Euglena* sp	4.6x10^{-2}
Gomphocybella sp	1.3x10^{-2}	*Trachelomonas* sp	2.6x10^{-2}
Chlorophyta		**Chlorophyta**	
Tetraedron trigonium	8.2x10^{-4}	*Coelastrum* sp	3.8x10^{-2}
		Oocystis lacutris	4.4x10^{-2}
Pond 4			
5cm depth			
Cyanobacteria			
Aphanocapsa sp	8.5x10^{-5}		
Planktolyngbya limnetica	8.7x10^{-4}		
Diatoms			
Epitheimia turgida	9.1x10^{-4}		
Fragilari a sp	1.9x10^{-4}		
Nitzschia acicularis	2.2x10^{-5}		
Euglenophytes			
Euglena sp	1.4x10^{-1}		
Phacus curvicuada	1.2x10^{-3}		
Chlorophyta			
Chlamydomonas sp	1.3x10^{-3}		
Coelastrum sp	1.8x10^{-2}		
45cm depth			
Cyanobacteria			
Planktolyngbya sp	8.3x10^{-3}		
Cryptophytes			
Cryptomonas sp	2.4x10^{-3}		
Diatoms			
Diatoma sp	1.5x10^{-2}		
Gomphocybella sp	2.4x10^{-2}		
Navicula sp	2.0x10^{-3}		
Nitzschia sp	3.7x10^{-3}		
Euglenophytes			
Trachelomonas sp	9.7x10^{-4}		
Chlorophyta			
Protoderma sp	4.5x10^{-3}		
Scenedesmus sp	4.9x10^{-3}		
70cm depth			
Cyanobacteria			
Merismopedia tenuissima	2.0x10^{-3}		
Planktolyngbya sp	5.4x10^{-1}		

Table A1 combined algal species list for period 1

Pond 1	Pond 2	Pond 3	Pond 4
Chlamydomonas sp	*Cryptomonas* sp	*Diatoma* sp	*Aphanocapsa* sp
Euglena acus	*Diatoma* sp	*Gomphocybella* sp	*Chlamydomonas* sp
Euglena sp	*Euglena* sp	*Gongosira* sp	*Coelastrum* sp
Gomphocybella sp	*Fragilaria* sp	*Planktolyngbya limnetica*	*Cryptomonas* sp
Merismopedia tenuissima	*Glenodinium* sp	*Planktolyngbya* sp	*Diatoma* sp
Nitzschia sp	*Gomphocybella* sp	*Planktothrix* sp	*Epitheimia turgida*
Planktolyngbya contarta	*Merismopedia tenuissima*	*Protoderma* sp	*Euglena* sp
Planktolyngbya sp	*Nitzschia acicularis*	*Tetraedron trigonium*	*Fragilaria* sp
Protoderma sp	*Planktolyngbya sp*		*Gomphocybella* sp
Strombomonas acuminatus	*Planktothrix* sp		*Merismopedia tenuissima*
Trachelomonas sp	*Protoderma* sp		*Navicula gastrum*
	Pseudodendoclonum sp		*Navicula* sp
	Scenedesmus sp		*Nitzschia acicularis*
	Tetraedron trigonium		*Nitzschia* sp
	Trachelomonas sp		*Oocystis lacutris*
			Phacus curvicuada
			Planktolyngbya limnetica
			Planktolyngbya sp
			Protoderma sp
			Scenedesmus sp
			Trachelomonas sp
Total 11	**15**	**8**	**21**

Table A2 combined algal species list for period 2

Pond 1	Pond 2	Pond 3	Pond 4
Aphanocapsa sp	*Aphanocapsa* sp	*Aphanocapsa* sp	*Aphanocapsa* sp
Cyclostephanodiscus	*Chroococcus dispersus*	*Cyclotella* sp	*Chroococcus* sp
Diatoma sp	*Cryptomonas* sp	*Diatoma* sp	*Cyclostephanodiscus*
Euglena sp	*Cyclostephanodiscus*	*Epitheimia* sp	*Diatoma* sp
Merismopedia tenuissima	*Diatoma* sp	*Fragilaria* sp	*Gomhocymbella* sp
Navicula sp	*Euglena* sp	*Merismopedia tenuissima*	*Merismopedia tenuissima*
Nitzschia acicularis	*Fragilaria* sp	*Navicula gastrum*	*Navicula gastrum*
Nitzschia sp	*Merismopedia tenuissima*	*Navicula* sp	*Navicula* sp
Phacus longicuada	*Navicula* sp	*Nitzschia acicularis*	*Nitzschia* sp
Planktolyngbya limnetica	*Nitzschia acicularis*	*Nitzschia* sp	*Phacus longicuada*
Planktolyngbya sp	*Nitzschia* sp	*Phacus longicuada*	*Planktolyngbya limnetica*
Protoderma sp	*Oocystis solitaria*	*Planktolyngbya limnetica*	*Planktolyngbya* sp
Rhodomonas sp	*Phacus longicuada*	*Planktolyngbya* sp	*Planktothrix* sp
Romeria graclie	*Planktolyngbya limnetica*	*Planktothrix* sp	*Protoderma* sp
Stuarastrum trigonum	*Planktothrix* sp	*Protoderma* sp	
Tetraedron trigonum	*Protoderma* sp	*Tetraedron trigonum*	
Trachelomonas sp	*Rhodomonas* sp	*Trachelomonas* sp	
	Rhopolodia sp		
	Scenedesmus sp		
	Tetraedron trigonum		
	Trachelomonas sp		
Total: 17	**21**	**17**	**14**

List 2 Algal composition on biofilm during period 2

Algal species	Biomass (gm^{-2})	Algal species	Biomass (gm^{-2})
Pond 1		**Euglenophytes**	
5cm depth		*Trachelomonas* sp	9.0×10^{-3}
Cyanobacteria		**Chlorophyta**	
Aphanocapsa sp	7.7×10^{-3}	*Protoderma* sp	7.3×10^{-2}
Merismopedia tenuissima	1.8×10^{-4}	**Pond 2**	
Planktolyngbya sp	9.5×10^{-2}	**5cm depth**	
Romeria graclie	4.0×10^{-4}	**Cyanobacteria**	
Cryptophytes		*Aphanocapsa* sp	1.5×10^{-2}
Rhodomonas sp	7.4×10^{-4}	*Chroococcus dispersus*	1.0×10^{-3}
Euglenophytes		*Merismopedia tenuissima*	4.4×10^{-4}
Euglena sp	1.9×10^{-1}	*Planktolyngbya limnetica*	3.3×10^{-2}
Phacus longicuada	3.5×10^{-1}	**Diatoms**	
Trachelomonas sp	6.4×10^{-3}	*Navicula* sp	2.1×10^{-1}
Chlorophyta		*Diatoma* sp	9.6×10^{-2}
Protoderma sp	6.2×10^{-2}	*Fragilaria* sp	1.2×10^{-2}
Stuarastrum trigonum	7.1×10^{-3}	*Nitzschia acicularis*	3.2×10^{-2}
Tetraedron trigonum	1.6×10^{-3}	*Nitzschia* sp	8.5×10^{-2}
Diatoms		**Euglenophytes**	
Diatoma sp	6.5×10^{-3}	*Euglena* sp	4.5×10^{-1}
Nitzschia sp	2.0×10^{-2}	*Trachelomonas* sp	5.8
		Chlorophyta	
45cm depth		*Protoderma* sp	3.7×10^{-1}
Cyanobacteria		*Scenedesmus* sp	2.2×10^{-2}
Aphanocapsa sp	1.0×10^{-2}	**45cm depth**	
Planktolyngbya limnetica	2.1×10^{-3}	**Cyanobacteria**	
Cryptophytes		*Aphanocapsa* sp	2.1×10^{-2}
Rhodomonas sp	2.7×10^{-3}	*Merismopedia tenuissima*	3.0×10^{-2}
Diatoms		*Planktolyngbya limnetica*	1.7
Diatoma sp	7.7×10^{-2}	**Diatoms**	
Navicula sp	1.5×10^{-2}	*Navicula* sp	2.0×10^{-2}
Nitzschia acicularis	1.2×10^{-3}	*Nitzschia* sp	1.1×10^{-1}
Euglenophytes		*Cyclostephanodiscus*	7.7×10^{-2}
Euglena sp	8.0×10^{-3}	*Diatoma* sp	6.7×10^{-2}
Trachelomonas sp	2.6×10^{-2}	*Rhopolodia* sp	2.9×10^{-2}
Chlorophyta		**Euglenophytes**	
Protoderma sp	9.2×10^{-1}	*Euglena* sp	5.3×10^{-1}
Tetraedron trigonum	8.4×10^{-4}	*Phacus longicuada*	7.3
		Trachelomonas sp	8.8×10^{-1}
70cm depth		**Chlorophyta**	
Cyanobacteria		*Oocystis solitaria*	3.0×10^{-2}
Aphanocapsa sp	1.4×10^{-2}	*Protoderma* sp	1.6×10^{-1}
Merismopedia tenuissima	6.5×10^{-5}	*Tetraedron trigonum*	6.3×10^{-3}
Planktolyngbya limnetica	9.9×10^{-3}	**70cm depth**	
Planktolyngbya sp	5.2×10^{-3}	**Cyanobacteria**	
Diatoms		*Aphanocapsa* sp	3.0×10^{-2}
Cyclostephanodiscus	8.8×10^{-3}	*Planktolyngbya limnetica*	1.1×10^{-1}
Diatoma sp	3.6×10^{-2}	*Planktothrix* sp	2.6×10^{-2}
Nitzschia acicularis	4.7×10^{-3}	**Cryptophytes**	
		Cryptomonas sp	6.4×10^{-3}
		Rhodomonas sp	3.6×10^{-3}

Continuation list 2, algal composition on biofilm during period 2

Algal species	Biomass (gm^{-2})	Algal species	Biomass (gm^{-2})
Diatoms		*Planktolyngbya limnetica*	$7x10^{-1}$
Cyclostephanodiscus	$2.2x10^{-2}$	*Planktothrix* sp	$4.9x10^{-2}$
Diatoma sp	$3.7x10^{-2}$	**Diatoms**	
Nitzschia sp	$5.1x10^{-2}$	*Diatoma* sp	$3.7x10^{-2}$
Rhopolodia sp	$5.7x10^{-3}$	*Navicula* sp	$5.4x10^{-2}$
Euglenophytes		*Nitzschia* sp	$4.1x10^{-2}$
Trachelomonas sp	$1.8x10^{-1}$	**Euglenophytes**	
Chlorophyta	$4.3x10^{-2}$	*Trachelomonas* sp	$1.7x10^{-1}$
Protoderma sp		**Chlorophyta**	
		Protoderma sp	$7.6x10^{-2}$
Pond 3		*Tetraedron trigonum*	$6.7x10^{-3}$
5cm depth			
Cyanobacteria	$3.0x10^{-2}$	**Pond 4**	
Aphanocapsa sp	5.2	**5cm depth**	
Planktolyngbya sp		**Cyanobacteria**	
Diatoms	$2.4x10^{-1}$	*Aphanocapsa* sp	$1.81x10^{-1}$
Nitzschia sp	$1.1x10^{-1}$	*Planktolyngbya limnetica*	$8.1x10^{-2}$
Diatom	$2.6x10^{-2}$	*Planktolyngbya* sp	4.3
Fragilaria sp	$4.9x10^{-2}$	*Planktothrix* sp	$6.8x10^{-1}$
Navicula gastrum		**Diatoms**	
Euglenophytes		*Navicula gastrum*	$8.4x10^{-3}$
Trachelomonas sp	$1.1x10^{-3}$	*Navicula* sp	$6.3x10^{-3}$
Chlorophyta		*Nitzschia* sp	1.2
Protoderma sp	2.1	**Chlorophyta**	
		Protoderma sp	2.2
45cm depth			
Cyanobacteria		**45cm depth**	
Planktothrix sp	$3.8x10^{-1}$	**Cyanobacteria**	
Aphanocapsa sp	$1.6x10^{-2}$	*Planktothrix* sp	4.0
Merismopedia tenuissima	$1.8x10^{-3}$	*Aphanocapsa* sp	$6.6x10^{-2}$
Planktolyngbya limnetica	$4.2x10^{-1}$	*Merismopedia tenuissima*	$1.5x10^{-2}$
Planktothrix sp	$7.3x10^{-2}$	*Planktolyngbya limnetica*	$1.8x10^{-1}$
Diatoms		**Diatoms**	
Cyclotella sp	$8.2x10^{-3}$	*Nitzschia* sp	$6.8x10^{-1}$
Diatoma sp	$3.7x10^{-2}$	*Diatoma* sp	$4.4x10^{-1}$
Epitheimia sp	$1.9x10^{-2}$	*Navicula* sp	$4.5x10^{-2}$
Navicula gastrum	$2.1x10^{-2}$	**Euglenophytes**	
Nitzschia acicularis	$4.3x10^{-2}$	*Phacus longicuada*	3.1
Nitzschia sp	$5.0x10^{-2}$	**Chlorophyta**	
Euglenophytes		*Protoderma* sp	1.4
Phacus longicuada	2.6		
Chlorophyta		**70 cm depth**	
Protoderma sp	$1.9x10^{-1}$	**Cyanobacteria**	
Tetraedron trigonum	$9.4x10^{-3}$	*Planktothrix* sp	$7.2x10^{-1}$
		Chroococcus sp	$1.2x10^{-2}$
70cm depth		*Merismopedia tenuissima*	$1.2x10^{-4}$
Cyanobacteria		*Planktolyngbya limnetica*	1.4
Aphanocapsa sp	$2.9x10^{-2}$	*Planktolyngbya* sp	2.0
Merismopedia tenuissima	$7.1x10^{-4}$		

Continuation list 2, algal composition on biofilm during period 2

Algal species	Biomass (gm^{-2})
Diatoms	
Cyclostephanodiscus	3.2×10^{-2}
Diatom	2.3×10^{-2}
Gomhocymbella sp	4.4×10^{-2}
Navicula gastrum	1.1×10^{-2}
Nitzschia sp	7.5×10^{-2}
Euglenophytes	
Phacus longicuada	1.4
Chlorophyta	
Protoderma sp	2.8×10^{-1}

Chapter 3
Comparison of hydraulic flow patterns of four pilot scale baffled wastewater stabilization ponds

Chapter 3

Comparison of hydraulic flow patterns of four pilot scale baffled waste stabilization ponds

Abstract
Four pilot scale wastewater stabilization ponds (WSP) were set up in Kampala – Uganda. Each pond had a length of 4m, width 1m and depth of 1m. The wastewater was filled up to a depth of 0.8m. Pond 1 was un-baffled and operated as control while ponds 2, 3 and 4 were fitted with fifteen baffles having the same surface area but different baffle configurations to induce different flow patterns. Tracer tests using lithium were performed, the aim of the test was to investigate the effect of baffles on the hydraulic characteristics of the ponds. The test was performed twice for all 4 ponds; the first test was run for 19 days while the second one lasted 30 days. The results for test 1 showed that the tracer concentration was not zero at the end time of the sample period; therefore a second run was performed. The tracer curves (lithium concentration-time curves) during the two runs looked similar, demonstrating the reproducibility of the test. Because the results for test 2 were more complete and therefore more reliable to use in the calculations, they are presented in the abstract. The tracer curves looked similar for ponds 1 and 2 implying that installing baffles parallel to the flow (as in pond 2) did not affect the flow pattern. The peak of the tracer curves of ponds 1 and 2 were reached first, followed by that of pond 3 and 4. The following parameters were calculated: the theoretical and actual (measured) mean hydraulic retention time; dead volume; short circuiting index, dispersion number and reactors in series. The theoretical mean hydraulic retention time were 6.2 days while the actual mean hydraulic retention times for ponds 1, 2, 3 and 4 were higher, 7.6, 7.5, 9.2 and 8.1 days. This can be explained by the pond design which resulted in the tracer diffusing in dead zones and being released slowly. This resulted in curves with long tails which gave higher actual mean hydraulic retention time than the theoretical mean hydraulic retention time. This resulted in negative dead volumes i.e. -23%,-21%,-49% and -60% for ponds 1, 2, 3 and 4 respectively. These negative values support the argument that the tracer diffuses in the dead volumes and is slowly released later. When comparing the ponds mutually, it was seen that the actual mean hydraulic retention times for ponds 3 and 4 for both tests were higher than those of ponds 1 and 2. This was believed to be due to the longer travel time and larger dead zones created by the baffle arrangements in ponds 3 and 4. The short circuiting index was 0.87 for ponds 1 and 2 and this decreases from pond 3 to 4. The baffle arrangement in pond 4 was effective in reducing short circuiting by 60%. According to both the mixers in series and dispersion model, pond 1 and 2 behaved like mixed reactors in series while pond 3 and 4 were best described by plug flow with moderate dispersion.

Key words: Tracer, stabilization ponds, baffles, dead volume, mean retention time

Introduction

The performance of wastewater stabilization ponds (WSP) is dependent on many factors; inter alia, organic loading regime, and geometry, climatic and environmental conditions. The hydraulic behavior of the ponds is also of principle importance in determining the overall treatment efficiency (Short *et al;* 2010). Understanding the hydraulics of WSP is vital in improving treatment performance (Nameche and Vasel, 1998; Shilton *et al.,* 2000).The objective of this study was to investigate the effect baffles on hydraulic characteristics of wastewater stabilization ponds under tropical conditions. Kilani and Ogunrombi (1984) and Muttamara and Puetpaiboon, (1997) studied the effect of baffles on nitrogen removal, but at laboratory scale. Knowledge on the hydraulics and use of baffles in ponds at pilot scale under tropical conditions is scanty.

Introduction of baffles in this research was considered with a dual purpose: (a) as support surface for biofilm development (Chapter 2), in particular for N-removal (Baskaran *et al.,* 1992; Craggs *et al.,* 2000; Mclean *et al.,* 2000) and (b) improvement of the hydraulic behavior (as in this chapter). However, introduction of baffles in ponds may affect the hydraulic characteristics of the ponds. For instance undesirable effects such as channelization, creation of dead zones and organic overloading (in-let area) have been reported (Shilton and Harrison, 2003). For ponds installed with horizontal baffles, 50% and 90% baffle-width relative to pond-width causes channeling and short-circuiting. It is recommended that a baffle width of 70% relative to pond width reduces channeling (Shilton and Harrison, 2003).

Generally, effects of baffles on pond hydraulics are not addressed by traditional pond design methods. It is difficult to reliably predict how different modifications and interventions affect pond performance (Wood *et al.,* 1995). Demonstrating hydrodynamic problems using Computational Fluid Dynamics (CFD) has been suggested. The major limitation is that computer programs are expensive and require expertise in application, and are focused on large scale systems. Levenspiel, (1972); Shilton *et al.,* (2000); Van der Steen, (2000); Zimmo, (2003) and many others applied tracers to study pond hydraulics and have found it useful. This approach was used in this study.

Theoretical background

In tracer studies, a tracer is introduced in the pond influent and its concentration at the effluent is determined in a series of grab samples collected at specific time intervals (Metcalf and Eddy, 2003). Tracer input in ponds is usually by two methods i.e. the step and pulse method. In the step method, the tracer is continuously added until the effluent concentration equals to influent concentration. In the pulse method, the tracer is added for a short time; usually the time of addition is shorter relative to the theoretical retention time.

There are a number of tracers that are used in scientific research but most common ones include Congo red, Fluorosilicic acid (H_2SiF_6), Hexafluoride gas, Lithium chloride (LiCl), Potassium permanganate, Rhodamine WT and Sodium chloride. Lithium chloride was used in this study due to its common application in studying natural systems. Besides, it can be easily analyzed using the atomic absorption spectrophotometer. Rhodamine WT is sensitive to light and temperature and requires a fluorometer, which is expensive equipment. Sodium chloride is the cheapest and easiest option but has a tendency of forming density currents unless mixed well (Metcalf and Eddy, 2003).

Tracer studies are important in assessing hydraulic characteristics of ponds. Normally, the time which water stays in the pond is calculated by dividing the pond volume and flow rate (Short et al., 2010). This is referred to as theoretical mean retention time and it assumes that the whole pond volume is active. The theoretical mean retention time usually differs from the actual (measured) mean retention time. This is caused by non-ideal flow in ponds which create pockets of stagnant water (Van der Steen, 2000). Levenspiel, (1972) and Agunwamba et al., (1992) describe reactors with non-ideal flow using the dispersion and mixers in series model. The dispersion model assumes plug flow which is superimposed on top with some degree of mixing. Depending on the intensity of intermixing, the prediction of this model ranges from plug flow at one extreme to mixed flow at the other end. Absence of intermixing represents an ideal plug flow situation while high mixing results in a completely mixed system (Metcalf and Eddy, 2003). The parameter used describing dispersion model is the dispersion number (d). The mixers in series model consider the system to be divided into a series of mixed reactor tanks. The parameter used to describe this model is the number of reactors in series (N). Since the behavior of the ponds was not known, both models were tested and the one which described the system approximately well was used.

The parameters (d) and (N) for both dispersion and the mixers in series model can be calculated from the tracer response curve using mean retention time and variance.

The mean retention time (t_m) is approximated if the concentration versus time tracer response curve (C) is defined by discrete time measurements as shown in equation 1 (Levenspiel, 1999).

$$t_m = \frac{\sum t_i C_i \Delta t_i}{\sum C_i \Delta t_i}$$

(1)

Where t_m = mean retention time based on discrete time step measurements (days)

   t_i = time at i^{th} measurement (days)

   C_i = tracer concentration i^{th} measurement (g m^{-3})

   Δt_i = time increment (days)

Variance σ^2 can be used to define the spread of the distribution (Levenspiel, 1999) as;

$$\sigma^2 = \frac{\int_0^\infty t_i^2 C_i(t_i) dt_i}{\int_0^\infty C_i(t_i) dt_i} - t_m^2$$

Or in discrete form as:

$$\sigma^2 = \frac{\sum t_i^2 C_i \Delta t_i}{\sum C_i \Delta t_i} - t_m^2 \tag{2}$$

Where σ^2 = variance based on discrete time measurements

Dispersion model
Dispersion number (d) is defined by Levenspiel (1972) as the coefficient of dispersion D divided by product of fluid velocity (u) and reactor length (L)

$$d = \frac{D}{uL} \tag{3}$$

Where d = dispersion number, (no units)

$\quad$ D = coefficient of dispersion (m^2s^{-1})

$\quad$ u = fluid velocity (ms^{-1})

$\quad$ L = length (m)

There is a relationship between variance of normalized tracer response $\sigma^2_{(\theta)}$, variance $\sigma^2_{(c)}$ derived from tracer response curve C, mean retention time t_m and dispersion number d for a pulse tracer input (Metcalf and Eddy, 2003) and this is given by:

$$\sigma^2_\theta = \frac{\sigma^2_c}{t_m^2} = 2\frac{D}{uL}$$

Hence $\quad 2d = \left(\frac{\sigma^2_c}{t_m^2}\right) \tag{4}$

Where σ^2_θ $\quad$ = variance of normalized tracer response C curve

$\quad\quad \sigma^2_c$ $\quad$ = variance derived from curve C (see equation 2)

$\quad\quad t_m$ $\quad$ = mean retention time

This implies that once the mean retention time and variance have been calculated from tracer results, dispersion number d can be calculated. The dispersion values can be used to assess the degree of axial dispersion in wastewater treatment. If d = 0, ideal plug flow; less than 0.05 - low dispersion; 0.05 to 0.25 – moderate dispersion; more than 0.25 high dispersion and when d tends to infinity, then the system is considered completely mixed (Metcalf and Eddy 2003).

Mixers in series model
The number of complete mixed reactors in series (N) can be calculated using the mean retention time and variance derived from tracer response curve C.

$$N = \frac{t_m^2}{\sigma_c^2} \qquad (5)$$

Where N = number of mixers in series.

The link between dispersion number (d) and number of mixers in series (N) can be established through the Peclet number (P_e). Peclet number is the inverse of dispersion number (d) and number of mixers in series (N) is a half the Peclet number (Metcalf and Eddy, 2003). For instance, 5, 3, 2.5, 2.0 and 1.7 mixed reactors in series will be required to simulate plug flow reactor with dispersion for dispersion numbers 0.1, 0.15, 0.20, 0.25 and 0.30 respectively.

Other parameters
After determination of mean theoretical retention time (t_{HRT}), the fraction of dead volume can be calculated as (Levenspiel, 1972):

$$Dead\ volume = \left(1 - \frac{t_m}{t_{HRT}}\right) \times 100\% \qquad (6)$$

Where t_{HRT} = theoretical hydraulic retention time (volume/flow rate)

The index of short circuiting (α_s) indicates how fast the influent reaches the effluent point. It is normally expressed as values that range from 0 to 1. When α_s approaches 1, the extent of short circuiting can be considered large.

$$\alpha_s = \frac{t_m - t_p}{t_m} \qquad (7)$$

Where

α_s = index of short circuiting
t_p = time taken to reach the maximum tracer concentration

Methodology
Description of pilot scale system
The pilot scale wastewater stabilization ponds used in this study was as described in chapter 2 (Figure 2). The operational conditions were also as described in chapter 2. After studying the effect of baffles on the biofilm formation and ecology of the wastewater stabilization ponds (Chapter 2), there was need to study the effect of baffles on the hydraulic characteristics of the ponds; this was addressed in this chapter.

Tracer experiment
The tracer experiment was performed twice. For the first test, lithium sulfate solution was used as the tracer. A solution of 21 liters containing 500 mg l^{-1} of lithium was prepared using tap water. This was fed into each maturation pond with a volume of 3200 liters; a maximum concentration of 3.3 mg l^{-1} of lithium was expected in the pond effluent (assuming it is a completely mixed system).

Four buckets each containing 21 liters of lithium solution were prepared as described above. The buckets were placed at the influent point of the maturation ponds and left for 2 hours to acclimatize to the environmental conditions. During this time, influent flow rates of the ponds were measured. The influent flow rates varied but a mean of 0.35 l min^{-1} was obtained for all ponds. After the 2 hours had elapsed, the influents of the ponds were closed, and the lithium from the buckets fed into the ponds at a flow rate of 0.35 l min^{-1}. The buckets were fitted with rubber stoppers connected to silicon tubes. At the end of each tube was a clip which was used in regulating the flows. After addition of lithium, the clips were closed and influent points of the ponds immediately opened and wastewater continued to flow at a rate 0.35l min $^{-1}$. Samples were taken immediately and for the next 19 days.

For the second test, LiCl.H$_2$O was used. It appeared that the lithium sulfate which was purchased at the local market and used for the first test was of insufficient quality. Therefore Lithium chloride, analytical grade (ACROS Organics, New Jersey) was obtained from UNESCO-IHE laboratories and used. Thirty liters of the effluent of FP was collected in plastic buckets and a lithium solution of 350 mg l^{-1} prepared. The buckets were put close to the maturation ponds and left for 2 hours to acclimatize to the environmental conditions. The solution was stirred and 25 liters added to each maturation pond at a rate of 0.36 l min^{-1} as described above. The experiment was run for 30 days because the first test showed that 19 days was not long enough. Samples were picked 3 times a day.

The addition of lithium to the ponds during the two tests can be considered as pulse input. This is because the addition took less than 1.5 hours as compared to the theoretical retention time of 149 hours (Metcalf and Eddy 2003).

Results
The results for the tracer studies during the two tests are presented in tables 3 and 3.1. Table 3 shows the amount of lithium added and the time it took for addition to each maturation pond. Table 3.1 shows the various parameters that are used to represent the hydraulic characteristics in ponds. The hydraulic characteristics were calculated based on Levenspiel, (1972, 1999) and Metcalf and Eddy, (2003) and the concentration-time data obtained as described above.

Table 3 Volumes and time which lithium was added to the maturation ponds

Test	Parameter	Volume added (l)	Duration (Hrs)
1	Pond 1	19	1.60
	Pond 2	19	1.00
	Pond 3	21	1.25
	Pond 4	21	1.25
2	Pond 1	25	1.14
	Pond 2	25	1.14
	Pond 3	25	1.14
	Pond 4	25	1.14

Table 3.1 Parameters used to describe hydraulic characteristics of the maturation ponds

Test	Parameter	Pond 1	Pond 2	Pond 3	Pond 4
1	Theoretical HRT (t_{HRT}, days)	6.3	6.2	6.2	6.4
	Actual HRT (t_m, days)	6.6	6.3	8.2	8.1
	Mixers in series (N)	1.8	1.8	2.7	4.2
	Short circuiting index (α_s)	0.88	0.88	0.66	0.29
	Dispersion (d)	0.28	0.28	0.19	0.12
	Dead volume (%)	-5.6	-2.0	-33	-26
	Recovery (%)	139	134	209	195
2	Theoretical HRT (t_{HRT}, days)	6.2	6.2	6.2	6.2
	Actual HRT (t_m, days)	7.6	7.5	9.2	9.9
	Mixers in series (N)	1.6	1.7	1.9	2.9
	Short circuiting index (α_s)	0.87	0.87	0.78	0.39
	Dispersion (d)	0.30	0.30	0.26	0.17
	Dead volume (%)	-23	-21	-49	-60
	Recovery (%)	78	77	101	94

Figure 3 and 3.1 show the normalized tracer concentration as function of normalized time. The results for both tests were similar and showed the curves to be skewed with long tails.

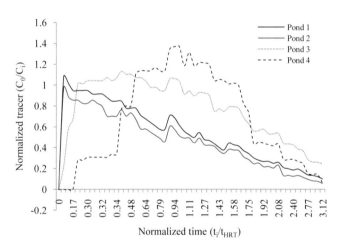

Figure 3: Normalized lithium concentration-time curves for the maturation ponds during test 1

considered completely mixed (see under section Dispersion model). Therefore, ponds 1 and 2 can be approximated to behave as ideal mixers. According to mixers in series model, pond 3 can be described by 2 ideally mixed reactors in series (Table 3.1). Two reactors in series corresponds to (d) of 0.25 which can be considered as plug flow with moderate dispersion. Pond 4 is described by 3 mixed reactors in series (Table 3.1) which corresponds to (d) of 0.15. The (d) value within 0.05 and 0.25 is also considered as plug flow with moderate dispersion (Metcalf and Eddy, 2003). According to Persson, (2000); the higher the number of stirred reactors in series (N), the more plug flow the system becomes. Table 3.1 shows N increasing from pond 1 to 4 probably suggesting more plug flow behavior. Therefore, ponds 3 and 4 are best described by plug flow with moderate dispersion. When the tracer results of test 1 were extrapolated, 1.4, 1.4, 1.9 and 3.2 mixed reactors in series were obtained for ponds 1, 2, 3 and 4 respectively. These results are similar to those obtained in test 2 where complete tracer curves were obtained indicating reproducibility of results.

Conclusions
The results of this study showed that the actual mean retention times based on the tracer study calculations for all the ponds during the two tests were higher than the theoretical mean retention times. This is probably caused by the tracer diffusing into dead volumes and slowly being released. It was also found that the un-baffled pond 1 and baffled pond 2 have similar hydraulic characteristics implying that installing baffles parallel to flow did not affect the pond hydraulics. However, the flow pattern of pond 3 and 4 were different (i.e. tracer peaks where reached at different times) indicating that the different baffle configurations were responsible for this observation. It can then be concluded that although addition of baffles can increase surface area for nitrifier growth, they have an effect on pond hydraulics depending on the type of configuration chosen. Additionally, baffle configuration in pond 1, 2 and 3 did not reduce short circuiting yet the baffle configuration of pond 4 was effective in reducing short circuiting. Ponds 1 and 2 behaved like mixed reactors while pond 3 and 4 were best described by plug flow with moderate dispersion.

References
Agunwamba, J.C., Egbuniwe, N. and Ademiluyi, J.O. (1992). Prediction of dispersion number in waste stabilization ponds. *Wat. Res. 26 (1), 85-89*

Baskaran, K., Scott, P.H. and Connor, M.A. (1992). Biofilms as an Aid to Nitrogen Removal in Sewage Treatment Lagoons. *Wat. Sci. Tech. 26(7-8), 1707-1716*

Bracho, N., Brissaud, F. and Vasel, J.L. (2009). Hydrodynamics of ponds Part ii practice. In: 8th IWA Specialist Group Conference on Waste Stabilization Ponds, 26 - 29 April 2009, Belo Horizonte/MG Brazil

Bracho, N., Lloyd, B. and Aldana, G. (2006). Optimization of hydraulic performance to maximize faecal coliform removal in maturation ponds. *Wat Res 40, 1677-1685*

Craggs, L.J., Tanner, C.C., Sukias, J.P.S. and Davies, C.R.J. (2000). Nitrification potential of attached biofilms in dairy wastewater stabilization ponds. *Wat .Sci. Tech.42 (10-11) 195-202*

Kilani, J.S. and Ogunrombi, J.A. (1984). Effects of baffles of the performance of model waste stabilization ponds. *Wat. Res. 18, 941-944*

Levenspiel, O., (1972). Chemical Reaction Engineering. Second edition, John Wiley and sons, New York

Levenspiel, O., (1999). Chemical Reaction Engineering. John Wiley and sons, New York

McLean B.M., Baskran, K., and Connor, M.A. (2000). The use of algal-bacterial biofilms to enhance nitrification rates in lagoons: Experience under laboratory and pilot scale conditions. *Wat .Sci. Tech. 42(10-11), 187-194*

Metcalf and Eddy, (2003). Wastewater engineering. Treatment and Reuse. Tchobanoglous, G., Burton, F.L., Stensel, H.D (Eds). 4th Ed. McGraw Hill, Inc., USA

Muttamara, S. and Puetpaiboon, U. (1997). Roles of Baffles in Waste Stabilization Ponds. *Wat. Sci. Tech. 35(8) 275-284*

Nameche, T.H and Vasel, J.L. (1998). Hydrodynamic studies and modelization for aerated lagoons and waste stabilization ponds. *Wat. Res. 32 (10). 3039-3045*

Perrson, J. (2000). The hydraulic performance of ponds of various layouts. *Urban Water 2, 243-250.*
Shilton, A. and Sweeney, D. (2005). Hydraulic design. In: Pond Treatment Technology (Ed). Shilton, IWA publishing, 189-217

Shilton, A., and Harrison, J. (2003). Guidelines for the Hydraulic design of waste stabilization ponds, Institute of technology and engineering, Massey University, New Zealand

Shilton, A., Wilks, T., Smyth, J. and Bickers, P. (2000). Tracer studies on a New Zealand waste stabilization pond and analysis of treatment efficiency. *Wat. Sci. Tech. 42 (10-11), 343-348*

Short, M.D., Cromar, N.J., and Fallowfield, H.J. (2010). Hydrodynamic performance of pilot-scale duckweed, algal-based, rock filter and attached-growth media reactors used for waste stabilization pond research. *Ecol. Eng. 36, 1700-1708.*

Thackston, E.L., Shields, F.D. and Schroeder, P.R. (1987). Residence time distributions of shallow basins. *J. Env. Engrg. Div. ASCE, 113: 1319-1332*

Torres, J.J., Soler, A., Saez, J and Llorens, M. (2000). Hydraulic performance of a deep stabilization pond fed at 3.5 m depth. *Wat. Res. 34 (3) 1042-1049*

Van der Steen, N.P. (2000). Fecal coliform removal from UASB effluent in integrated systems of algae and duckweed. PhD Thesis, Ben-Gurion University of Negev, Israel

Wood, M.G., Greenfield, P.F., Howes, T., Johns, M.R., Keller, J. (1995). Computational Fluid Dynamic Modeling of Wastewater Ponds to Improve Design. *Wat. Sci. Tech. 31(12), 111-118*

Zimmo, O.R. (2003). Nitrogen transformations and removal mechanisms in algal and duckweed waste stabilization ponds. PhD Dissertation, UNESCO-IHE, Wageningen University, Netherlands

Chapter 4
Nitrification in bulk water and biofilms of algae wastewater stabilization ponds

Published as: Nitrification in bulk water and biofilms of algae wastewater stabilization ponds. M.A. Babu, M.M. Mushi, N.P van der Steen, C.M. Hooijmans and H.J. Gijzen. Wat. Sci. Tech. 55 (11), 93-101, 2007.

Chapter 4

Nitrification in bulk water and biofilms of algae wastewater stabilization ponds

Abstract

Nitrogen removal in wastewater stabilization ponds is poorly understood and effluent monitoring data show a wide range of differences in ammonium. For effluent discharge into the environment, low levels of nitrogen are recommended. Nitrification is limiting in facultative wastewater stabilization ponds. The reason why nitrification is considered to be limiting is attributed to low growth rate and wash out of the nitrifiers. Therefore to maintain a population, attached growth is required. The aim of this research was to study the relative contribution of bulk water and biofilms with respect to nitrification. The hypothesis was that nitrification can be enhanced in wastewater stabilization ponds by increasing the surface area for nitrifier attachment. In order to achieve this, transparent pond reactors representing water columns in algae WSP were used. To discriminate between bulk and biofilm activity, 5 day-batch activity tests were carried out with bulk water and biofilm sampled from the pond reactors. The observed value for $R_{bulk5day}$ was 2.7×10^{-4} g-N l^{-1} d^{-1} and for $R_{biofilm}$ was 1.50 g-N m^{-2} d^{-1}. During the 5 days of experiment with the biofilm, ammonia reduction was rapid on the first day. Therefore, a short-term biofilm activity test was performed to confirm this rapid decrease. Results revealed a nitrification rate, $R_{biofilm}$, of 2.13-N m^{-2} d^{-1} for the first 5 hours of the test, which was significantly higher than the 1.50 g-N m^{-2} d^{-1}, observed on the first day of the 7-day biofilm activity test. Results of this study demonstrated that biofilm nitrification rates were significantly higher than the bulk water nitrification rates although oxygen concentration in the latter was kept high at 8.8 mg l^{-1}. This implies that biofilms could play an important role in improving nitrification process in wastewater stabilization ponds. The volatilization rates were low even for experiments were air was bubbled to keep the oxygen concentration high.

Key Words: Nitrification, Wastewater Stabilization Ponds, Bulk water, Biofilm,

Introduction

Nitrogen pollution on the world water bodies is increasing and effects have become visible since the 1960's when increasingly reports are given of eutrophication of water bodies (Gijzen and Mulder, 2001). In response, high environmental standards and stringent regulations are being adopted by developed nations. Many developing countries have followed suit and have set strict standards, which in practice do not function because of prohibitive costs for treatment plants required to satisfy those standards (Veenstra and Alaerts, 1996). Several advanced treatment technological innovations have come up but developing nations cannot afford them, yet urbanization and population is on the rise (Gijzen and Khonker, 1997; Yu *et al.*, 1997; Gijzen *et al.*, 2004). This is compounded further by the millennium development goal seven which advocates for reduction of half the proportion of people without access to safe drinking water by 2015 (WSSCC,2004). Increase in accessibility to safe drinking water and sanitation implies generation of more wastewater. It is estimated that 80-90% of water consumed is converted to wastewater (Mara *et al.*, 1992).

In trying to address the issues of wastewater treatment, most developing countries have opted for wastewater stabilization ponds (WSP) as the major treatment technology. This is due to its cost-effectiveness in construction and maintenance. In fact Mara and Pearson, (1998) recommend the use WSP in developing regions. They have found them to perform similar to advanced systems especially in COD removal. However, their major shortcomings includes narrow zone for nitrification since the aerobic zone is limited to the upper 0.50 m (Baskran *et al.*, 1992); long hydraulic retention time and low attached bacterial biomass (McLean *et al.*, 2000; Zimmo *et al.*, 2000); short-circuiting (Shilton *et al.*, 2000; Shilton and Harrison, 2003); high concentration of total suspended solids (TSS) in the effluent (Mara, 2004); and large area for construction (Pearson, 1996).

This study mainly focused on limitation of nitrification due to lack of attachment surface for nitrifiers. In this study, nitrification rates in the bulk water and in the biofilm were investigated through a series of batch activity tests.

Materials and methods
Experimental set-up
Four transparent pond reactors with a surface area of 0.043 m^2, an effective depth of 0.95 meter and a volume of 0.041 m^3 were used. The pond reactors A1-A3 were placed in series, while A0 was a single reactor. The reactors simulated the water column in algae wastewater stabilization ponds (Figure 4.1). Synthetic wastewater (Table 4.1) of moderate strength (Metcalf and Eddy, 2003) was continuously fed at a flow rate of 0.66 l d^{-1} into A0 and A1, which translated into a theoretical mean retention time of 2.6 days for each pond reactor. Nitrifiers and denitrifiers were introduced into the system at the start of the experiment by seeding with 100 ml of aerobically and anoxically grown activated sludge. Later on, green algae were also introduced in the reactors and allowed to colonize the system. The set-up was exposed to 12-hour light and dark regimes by illumination with a light intensity of 125-129 $\mu Em^{-2}s^{-1}$. This provided sufficient light and represented natural conditions. The lamps also provided heat that resulted in a mean ambient temperature of 24°C. The average influent NH_4-N and COD concentration was 40 mg l^{-1} NH_4-N and 96 mg l^{-1} COD respectively.

Bulk water activity tests
One liter of bulk water was collected from A3 and placed into two 2-liter glass beakers (horizontal surface area 0.0177 m^2). Similarly, synthetic wastewater (Table 4.1) devoid of microorganisms was prepared with an ammonia concentration adjusted to 20 mg l^{-1} - values similar to that of the effluent bulk water of A3 and placed into two beakers to serve as control. All beakers were continuously aerated and exposed to light for 7 hours. Water samples were collected on an hourly basis and the ammonium concentration determined according to standard methods (APHA, 1995). The experiment was continued for 7 days. Ammonia, pH and oxygen levels were monitored on a daily basis (APHA, 1995). All the experiments were run in duplicates.

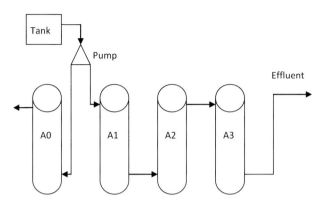

Figure 4.1: Experimental set up showing flow patterns of A0, A1, A2 & A3

Table 4.1 Composition of synthetic wastewater (modified after Moussa *et al.,* 2003)

Macro nutrients	Concentration (mg/l)	Micro nutrients solution	Concentration (g/l)
CH_3OONH_4	93.75	EDTA	10
NH_4Cl	87.70	$FeCl_3.6H_2O$	1.5
$NaH_2PO_4.H_2O$	26.70	H_3BO_3	0.15
$MgSO_4.7H_2O$	90.00	$CuSO_4.2H_2O$	0.03
$CaCl_2.2H_2O$	4.72	KI	0.18
KCl	36.00	$MnCl_2.4H_2O$	0.12
Micronutrient solution	0.6 (ml/l)	$Na_2MoO_4.2H_2O$	0.06
		$ZnSO_4.7H_2O$	0.12
		$CoCl.6H_2O$	0.15

Biofilm activity test (7 days duration)

Algal biofilm was carefully collected from the walls of A3 at 0.05m depth. A small round sampler of an area of 2.7×10^{-3} m^2 was found to collect an average of 3.30g ± 0.47 wet weight of biofilm. Biofilm of 11.28g (equivalent to biofilm area of 0.0092 m^2) and 11.30g (0.0098 m^2) were sampled for duplicate studies. These were placed in beakers (area 0.0177 m^2) containing 1 liter of synthetic wastewater (ammonium concentration of 20 mg l^{-1}) devoid of microorganisms. The biofilms were carefully handled to avoid disintegration. Control experiments using the same synthetic wastewater devoid of microorganisms and biofilm were run in parallel. All the beakers were continuously exposed to 12-hour light/dark regimes. The experiment was left to run for 7 days without aeration. Ammonia, pH and oxygen were monitored according to APHA (1995) on a daily basis.

Biofilm activity test (7 hours duration)

After the 7 days duration of the experiment above, the biofilms were reused to run a short-term experiment. At this time, the biofilm structure was still intact. This experiment lasted for a period of seven hours. The biofilms were weighed and the wet weight had increased from 11.28g to 12g (0.00981 m^2) and 11.30g to 13.73g (0.01123 m^2) respectively. The same experimental procedures as above were repeated. Water samples were collected hourly and ammonia concentration, pH and oxygen determined (APHA, 1995).

Biofilm plate activity tests-(7 hours duration)
To investigate further nitrification rates of biofilm from A3, glass biofilm plates of 0.03m by 0.08m were suspended in this pond reactor at 0.05m depth (Van der Steen, 2000). The plates were retrieved from the pond reactor after two weeks and hung in glass beakers containing 0.5 liters of synthetic wastewater containing 20 mg l^{-1} NH$_4$-N. Ammonium nitrogen, nitrates, pH, dissolved oxygen and temperature were monitored (APHA, 1995) after every two hours.

Ammonia Volatilization
Ammonia volatilization was calculated using Equation (1) developed by Zimmo *et al.,* (2004):

Ammonia volatilization rate (g-N m^{-2} d^{-1})   = 3.3[NH$_3$-N] +4.90, (1)

Where, [NH$_3$-N] is calculated from Emerson *et al.,* (1975) as:

$$\% \; Unionised \; NH_3 \;\; = \frac{100}{1+10^{(pKa-pH)}} ,$$ (2)

The temperature and pH measured during the activity tests were used to calculate percentage unionized ammonia which was used in equation 1 to calculate ammonia volatilization rates.

Statistical analysis
For statistical analysis, the ammonia concentration was plotted against time and the linear parts of the slopes of different experiments compared. For experiments which were run for days, the time was converted to hours before the graphs were plotted. Regression analysis using the F-test (95% confidence interval, at 0.05 levels) was used to check if the slopes for different treatments were statistically different.

Results
The results for bulk water experiments are shown in figures 4.2 and 4.3. The 7 hour bulk water test did not show any decrease in ammonia (Figure 4.2). A mean ammonia concentration of 21.1 ± 0.6 mg l^{-1} NH$_4$-N and 19.6 ± 1.1 mg l^{-1} NH$_4$-N was measured in the duplicate experiment 1 and 2, respectively. The control experiment showed a similar trend with a mean ammonia concentration of 20-21 mg l^{-1} NH$_4$-N. However, when the experimental time of the bulk water test was increased to 5 days, there was a slight drop of ammonia in the first four days followed by a rapid drop there after (Figure 4.3). The concentrations dropped from 20.1 to 17.4 mg l^{-1} NH$_4$-N and from 19.6 to 16.8 mg l^{-1} NH$_4$-N in duplicate 1 and 2, respectively. After the fourth day, the ammonia concentration rapidly dropped to 9.5 mg l^{-1} NH$_4$-N giving an overall ammonia decrease of 10.6 mg l^{-1} NH$_4$-N for the entire period of five days (Figure 4.3).

growth of nitrifiers. Nitrifiers are known to be slow growers thus would require more time to build their population.

Like in the bulk water, there was rapid drop of the ammonia concentration in the control after four days. The overall ammonia decrease from the control was 6.5 and 8.7 mg l^{-1} NH$_4$-N in control 1 and 2 respectively. These values are slightly lower than the values from the bulk water. Ammonia loss in the control is exclusively due to volatilization while in the bulk water, both ammonia oxidation and volatilization may have been important. Taking values of 10.6 mg l^{-1} NH$_4$-N and 8.7 mg l^{-1} NH$_4$-N as decrease in bulk water and control, it can be assumed that 1.9 mg l^{-1} NH$_4$-N was lost due to nitrification. This gave a bulk water nitrification rate ($R_{bulk5day}$) of 2.7×10^{-4} g-N l^{-1} d^{-1}. The result indicated very low nitrification rates in the bulk water. This is in agreement with McLean *et al.*, (2000) who made a similar observation for algae lagoons. They observed that lagoons with a high density of suspended algae had higher nitrification rates. The algae provided attachment surface for nitrifier growth since they are known to prefer attached growth. The decrease of ammonia in the control showed that long term bubbling of air strips the water of ammonia.

Biofilm activity test, 7 days experiment
The ammonium concentration in the biofilm experiment rapidly dropped from 18.0 to 3.8 mg l^{-1} NH$_4$-N within one day (Figure 4.4). The concentrations then gradually dropped and after the fifth day, very low concentrations (< 1 mg l^{-1} NH$_4$-N) were measured. While the ammonia concentrations dropped, nitrate levels built up (Figure 4.4) and became constant after 3 days. However, the amount of nitrate that built up was lower than the ammonia reduced, showing that nitrates did not accumulate. It is believed that nitrification and denitrification occurred simultaneously, with denitrification occurring in the deeper anoxic micro environments of the biofilm (Kuenen and Robertson, 1994). This could have resulted in the low nitrate concentration. Presence of nitrates; although at lower concentrations was clear evidence that nitrification occurred in the beakers (Pearson, 2005).

Biofilm nitrification rate ($R_{biofilm}$) of 1.50 g-N m^{-2} d^{-1} was obtained for this experiment which was comparable to 1.65 g-N m^{-2} d^{-1} obtained for the glass biofilm plates (Table 4.2). This showed that the biofilm glass plates could be used to determine nitrification rates of the pond reactors. In comparison to the biofilm experiment, the ammonia reduction in the control experiment was small i.e. from 26.5 to 22.6 mg l^{-1} NH$_4$-N with an average ammonia concentration of 23.0 ± 2.8 mg l^{-1} NH$_4$-N obtained after seven days of incubation (Figure 4.4). This showed that ammonia volatilization was minimal. Ammonia volatilization rate of 5.9×10^{-3} g-N m^{-2} d^{-1} was obtained for the duplicate control experiments (Table 4.2). The slope of the control experiment (-0.21) was significantly different from those of biofilm experiment (-0.7) (df = 2, 3; F = 29.9) implying that the biofilm nitrification rates were significantly higher than the volatilization rates. It can then be concluded that ammonia loss in the biofilm experiment was mostly due to nitrification. The results of biofilm experiment in figure 4.4 showed a rapid decrease of ammonia during the first day, a short term experiment was conducted to investigate this further and results are discussed below.

Biofilm activity test, 7 hour- experiments
A short-term biofilm activity test was performed to confirm this rapid decrease. Both duplicates 1 and 2 containing biofilm samples showed a gradual drop in ammonia concentration with time (Figure 4.5). A regression line was fitted and ammonia reduction rate of 0.85 and 0.83 mg l^{-1} h^{-1} (R^2= 0.93 and 0.86) were obtained for duplicate 1 and 2 respectively. The average values of these reduction rates gave a biofilm nitrification rate, ($R_{biofilm}$) of 2.13 g-N m^{-2} d^{-1} for the first 5 hours of the test, which was significantly higher (df = 2,7; F= 5.2) than the 1.50 g-N m^{-2} d^{-1}, observed on the first day of the seven-day biofilm activity test (Table 4.2). The results for the short term experiment are more reliable since the experiment was closely monitored. Both the $R_{biofilm}$ values for short and long term experiments are comparable to 0.72-2.64 g-N m^{-2} d^{-1} (Leu *et al.*, 1998) but higher than 0.48-0.72 g-N m^{-2} d^{-1} (Craggs *et al.*, 2000) and 0.72-0.96 g-N m^{-2} d^{-1} (McLean *et al.*, 2000). The control experiment showed a constant ammonia concentration during the first five hours of the experimental period. This again confirmed that volatilization was minimal. These results are in agreement to those obtained by Zimmo *et al.*, (2004), which indicated that ammonia volatilization was negligible in wastewater stabilization ponds under these conditions.

Conclusions
This study demonstrated the importance of attached growth in the process of improving nitrification in wastewater stabilization ponds. The results showed that the biofilm nitrification rates were significantly higher than bulk water nitrification rates. The volatilization rates were low and probably play a negligible role in wastewater stabilization ponds. Nitrates were found to accumulate in biofilm experiments although the accumulation did not equal the concentration of ammonia reduced. Presence of nitrates indicated nitrification process and the lower concentrations relative to amount of ammonia reduced could be an indication of denitrification.

Acknowledgements
We are grateful for the support from the Netherlands government through Netherlands Fellowship Program and the EU-Switch project contract number 60030361 for financial assistance. The authors' are also thankful to Edwin Hes, Shi Wenxin and UNESCO-IHE laboratory staff for their assistance and support in the laboratory work.

References
APHA. (1995). Standard Methods for Examination of Water and Wastewater 19th Ed., Washington, D.C

Baskaran, K., Scott, P.H., and Connor, M.A. (1992). Biofilms as an Aid to Nitrogen Removal in Sewage Treatment Lagoons. *Wat. Sci. Tech.* **26***(7-8), 1707-1716*

Craggs, L.J., Tanner, C.C., Sukias, J.P.S and Davies, C.R.J. (2000). Nitrification potential of attached biofilms in dairy wastewater stabilization ponds. *Wat. Sci. Tech.* **42***(10-11), 195-202*

Emerson, K., Russo, R.E., Lund, R.E., Thurston, R.V. (1975). Aqueous Ammonia Equilibrium Calculations: Effect of pH and Temperature. *Jour. Fisheries Res. Board of Canada* **32** *(12) 2379-2383*

Gijzen, H.J., Bos J.J., Hilderink, H.B.M., Moussa M., Niessen L.W., and de Ruyter van Stevennck E.D. (2004): Quick scan health benefits and costs of water supply and sanitation. Netherlands Environmental Assessment Agency. National Institute for Public Health and the Environment – (MNP-RIVM). The Netherlands

Gijzen, H.J., and Khondker, M. (1997). An overview of ecology, physiology, cultivation and application of duckweed, Literature review. Report of Duckweed Research project. Dhaka, Bangladesh.

Gijzen, H.J and Mulder, A. (2001). The global nitrogen cycle out of balance. *Water* **21**, Aug 2001, 38-40

Hammer, D.A., and Knight, R.L. (1994). Designing constructed wetlands for nitrogen removal. *Wat. Sci. Tech.* **29***(4), 15-27*

Kuenen, G.J and Robertson, L.A. (1994).Combined nitrification-denitrification processes. *FEMS Microbiology reviews* **15** *(1994) 109-117*

Leu, H.G., Lee, C.D., Ouyang, C.F. and Tseng, H. (1998). Effects of organic matter conversion rates of nitrogenous compounds in a channel reactor under various flow conditions. *Wat, Res. Vol. N.* **3***, 891-899*

Mara, D.D. (2004). Domestic wastewater treatment in developing countries. Earth scan, London

Mara, D.D., and Pearson, H.W. (1998). Design manual for waste stabilization ponds in Mediterranean countries. European Investment bank. Lagoon Technology International Ltd Leeds, England

Mara, D.D., Alabster, G.P., Pearson, H.W and Mills, S.W. (1992). Waste stabilization ponds, a design manual for Eastern Africa, Lagoon Technology International Leeds, England

McLean, B.M., Baskran, K., and Connor, M.A. (2000). The use of algal-bacterial biofilms to enhance nitrification rates in lagoons: Experience under laboratory and pilot scale conditions. *Wat .Sci. Tech.* **42(10-11), 187-194**

Metcalf and Eddy (2003). Wastewater engineering. Treatment and Reuse. Tchobanoglous, G., Burton, F.L., Stensel, H.D (Eds). 4th Ed. McGraw Hill, Inc., USA

Metcalf and Eddy (1991). Wastewater engineering. Treatment, Disposal and Reuse, 2nd Ed. Revised by Tchobanoglous, G., Burton, F.L. McGraw Hill, Inc., USA

Moussa, M.S., Lubberding, H.J., Hooijmans, C.M., van Loosdrecht, M.C.M., Gijzen, H.J. (2003). Improved method for determination of ammonia and nitrite oxidation activities in mixed bacterial culture. *Appl. Microbiol Biotechnol.* **63***: 217-221*

Pearson, H.W. (2005). Microbiology of waste stabilization ponds. In: Pond Treatment Technology (Ed). Shilton, IWA publishing

Pearson, H.W. (1996). Expanding the horizons of pond technology and application in an environmentally conscious world. *Wat. Sci. Tech.* **33(7)***, 1-9*

Shilton, A., and Harrison, J. (2003). Guidelines for the Hydraulic design of waste stabilization ponds, Institute of technology and engineering, Massey University, New Zealand

Shilton, A., Wilks, T., Smyth, J., and Bickers, P. (2000). Tracer studies on a New Zealand waste stabilization pond and analysis of treatment efficiency. *Wat. Sci. Tech* **42(10-11)***, 343-348*

Van der Steen, N.P. (2000): Faecal coliform removal from UASB effluent in integrated systems of algae and duckweed. PhD Thesis, Ben-Gurion University of Negev, Israel.

Veenstra, S. & Alaerts, G. (1996). Technology selection for pollution control. In: A. Balkema, H. Aalbers and E. Heijndermans (Eds.), Workshop on sustainable municipal waste water treatment systems, Leusden, the Netherlands, 17-40

WSSCC (2004). Resource packs on the water and sanitation Millennium development Goals. Water supply and sanitation collaborative council, Geneva

Yu, H., Tay, J., Wilson, F. (1997). A sustainable municipal wastewater treatment process for tropical and subtropical regions in developing countries. *Wat. Sci. Tech.* **35 (9)***, 191-198*

Zimmo, O.R, van der Steen N.P., and Gijzen H.J. (2004). Nitrogen mass balance across a pilot-scale algae and duckweed based wastewater stabilization ponds. *Wat. Res. Vol* **38(4)***, 913-920*

Zimmo, O.R. (2003). Nitrogen transformations and removal mechanisms in Algal and Duckweed Wastewater Stabilization Ponds. PhD Dissertation, and UNESCO - IHE institute for water education Wageningen University, The Netherlands

Zimmo O.R., Al sa'ed R., Gijzen, H.J. (2000). Comparison between algae based and duckweed based wastewater treatment. Differences in environmental conditions and nitrogen transformations. *Wat. Sci. Tech. 42(10-11), 215-222*

Chapter 5
Nitrification rates of algal-bacterial biofilms in wastewater stabilization ponds under light and dark conditions

Published as:
M.A. Babu., E.M.A. Hes., N.P. van der Steen., C.M. Hooijmans and H.J. Gijzen. Nitrification rates of algal-bacterial biofilms in wastewater stabilization ponds under light and dark conditions. Ecological Engineering 36 (2010) 1741–1746

Chapter 5

Nitrification rates of algal-bacterial biofilms in wastewater stabilization ponds under light and dark conditions

Abstract

The objective of this study was to investigate nitrification rates in algal-bacterial biofilms of wastewater stabilization ponds (WSP) under different conditions of light, oxygen and pH. Biofilms were grown on wooden plates of 6.0 cm by 8.0 cm by 0.4cm in a PVC tray continuously fed with synthetic wastewater with initial NH_4-N and COD concentrations of 40 mg l^{-1} and 100 mg l^{-1} respectively, under light intensity of 85-95 $\mu Em^{-2}s^{-1}$. Batch activity tests were carried out by exposure of the plates to light conditions as above (to simulate day time), dim light of 1.8- 2.2 $\mu Em^{-2}s^{-1}$ (to simulate reduced light as in deeper locations in WSP) and dark conditions (to simulate night time). Dissolved oxygen concentration and pH were controlled. At some experiments both parameters were kept constant and at others left to vary as in WSP. Results showed biofilm nitrification rates of 0.95-1.82 g-$Nm^{-2}d^{-1}$ and 0.16-1.62 g-$Nm^{-2}d^{-1}$ for light and dark experiments. When the minimum DO was 4.1 mg l^{-1}, the biofilm nitrification rates under light and dark conditions did not differ significantly at 95% confidence. When the minimum DO in the dim light experiment was 3.2 mg l^{-1}, the nitrification rates under light and dim light conditions were 0.95 g-$Nm^{-2}d^{-1}$ and 0.56 g-$Nm^{-2}d^{-1}$ and these significantly differed. Further decrease of DO to 1.1 mg l^{-1} under dark conditions resulted in more decrease of the nitrification rates to 0.16 g-$Nm^{-2}d^{-1}$. It therefore seems that under these experimental conditions, biofilm nitrification rates are significantly reduced at certain point when bulk water DO is between 3.2 and 4.1 mg l^{-1}. As long as bulk water DO under dark is high, light was not important in influencing the process of nitrification.

Key words: Biofilm, nitrification, light intensity, WSP, oxygen, pH

Introduction

Wastewater stabilization ponds are used worldwide as an effective and low cost technology for wastewater treatment. The major disadvantage of ponds is the large land area required for construction and their limited efficiency in removing nitrogen. They are usually designed as a sequence of systems (anaerobic reactor or pond, facultative ponds and maturation ponds). It is suggested that biofilms could be introduced into maturation ponds to reduce the required land area (Johnson and Mara, 2005; Xia *et al.*, 2008) and improving nitrogen removal. Introduction of biofilms into WSP has been investigated by McLean *et al.*, (2000) and seems to be effective in increasing nitrification rates but the effect of typical variations in the pond environment (pH, temperature, light intensity, oxygen concentration) on algal-bacterial biofilm nitrification is largely unknown. Various experiments on nitrification conducted by Zimmo *et al.*, (2004) were limited to bulk water nitrification. Information on nitrification rates of algal-bacterial biofilms of WSP under different conditions is still insufficient. Studies by Wolf *et al.*, (2007) and Roeselers *et al.*, (2008) investigated phototrophic biofilms and suggest that biofilm systems have a future prospect in wastewater treatment. Through modeling, Wolf *et al.*, (2007) have shown enhanced oxygen production within the phototrophic biofilms under light conditions and this could be used to the advantage of the nitrifiers. This study was conducted at laboratories of UNESCO-IHE, Delft and is part of other studies on pilot scale wastewater stabilization ponds carried out in Uganda. The focus of this study was to investigate biofilm nitrification rates under light and dark conditions as well as under different pH and oxygen concentrations.

Methodology
Growth of biofilm
A PVC tray (55cm x 37cm x 9.0 cm; length, width and depth) was divided into 3 compartments by two equally spaced transparent acrylic plates (Figure 5). The compartments were connected by openings at each opposite end. Sixty wooden biofilm plates of 6.0 by 8.0 cm were suspended vertically and parallel to the flow. Wood was chosen because it provides a rough surface which is thought to improve biofilm attachment.

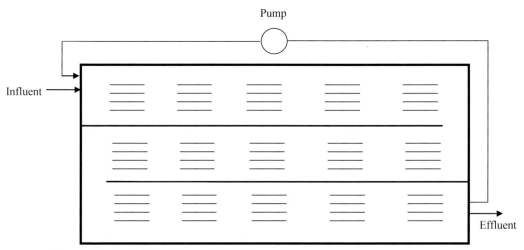

Figure 5: Experimental set up for growth of biofilm

The system was continuously fed with synthetic wastewater of ammonia and COD concentration of 40 mg N l^{-1} and 100 mg l^{-1}, respectively (Babu et al., 2007). The influent flow rate was 0.96 l hr^{-1} and the effluent was recycled at a rate of 2.5 l hr^{-1} just before the final outlet to ensure a uniform distribution of ammonia over the reactor. Enriched activated sludge (100 ml) from Hoek van Holland municipal wastewater treatment plant was used as inoculum to establish nitrifier and denitrifier populations. Algae (100ml) from the column experiments of Babu et al., (2007) were introduced into the system. The set-up was exposed to a 12 hour light regime of 85-95 $\mu Em^{-2} s^{-1}$. The plates were left to develop biofilms for about 60 days and then transferred to batch reactors for determination of the nitrification rates. The same synthetic wastewater was used for the continuous flow and batch systems.

Batch reactors for nitrification activity
The purpose of these experimental runs was to identify the effect of light and oxygen on biofilm nitrification rates. Batch experiments were conducted in duplicates under different conditions of light and oxygen and nitrification rates were determined in the laboratory using batch tests as described below:

1. Light and dark conditions; oxygen and pH left to vary
Biofilm plates from the tray were removed, gently rinsed with distilled water and hung vertically in two-liter glass beakers containing 1.1 l of fresh synthetic wastewater (not seeded with nitrifiers) of ammonia concentration of 20 mg l^{-1} and COD of 100 mg l^{-1} (Babu *et al.*, 2007). Each beaker had one biofilm plate which was exposed to a light intensity of 85-95 µEm^{-2}s^{-1} for a period of 8 hrs; this was lower than 133-176 µEm^{-2}s^{-1} measured in Uganda on a sunny day. Oxygen and pH were left to vary as in WSP. The temperature was not controlled but was almost constant (22°C). Samples were taken after every two hours, filtered and ammonia measured. Other parameters monitored included nitrate, DO, pH and temperature. All parameters were determined according to APHA, (1995). The procedures above were repeated but in this case, the beakers with biofilm were exposed to dark conditions. Similarly, oxygen and pH were left to vary as in WSP. A control with only synthetic wastewater without biofilm plates was also set up and exposed to light as above. This was to determine ammonia loss by volatilization.

2. Light and dark conditions; oxygen and pH kept constant
The experiments as in (1) above were repeated using 0.5 liters of synthetic wastewater with oxygen and pH kept constant. The DO was kept between 9.5 and 6.7 mg l^{-1} by continuous bubbling of air while the pH was kept at 7.7 by addition of 150 mg l^{-1} of sodium bicarbonate to the synthetic wastewater. Control experiments with only synthetic wastewater with pH and oxygen conditions similar to biofilm experiments were also run.

3. Light and dark conditions; pH kept constant but oxygen left to vary
In this experiment, the pH was kept constant as described above but the oxygen concentrations were left to vary with time. Control experiments without biofilm plates were also run under these conditions.

4. Light, dim light and dark conditions; pH kept constant, oxygen kept constant as per condition
The experiments above were repeated but this time under the conditions of bright light, dim light and darkness. Bright light represented the top part of the WSP at day light while dim light simulated the deeper and shaded parts of WSP during day. Dark conditions were to simulate night time. The oxygen level under bright light was kept between 8-9 mg l^{-1} by continuous bubbling with air. The oxygen level under dim light was kept between 3-5 mg l^{-1} by periodic bubbling with air, while that under dark conditions was kept between 2.3-1.1 mg l^{-1} by periodic bubbling with air and nitrogen gas. The pH was kept constant as

described above. This experiment was run for 6 hours and not 8 hours as the others. Also, 0.5 liters of synthetic wastewater was used.

Algal uptake

Since ammonia removal is not only due to nitrification but also by algal growth, the nitrification rates under light conditions were corrected for the nitrogen uptake by algae. The data for biofilm biomass (dry weight) taken from the pilot scale studies (Chapter 2) were used for correction. It was assumed that the ammonia uptake rates by algae in the pilot scale and laboratory and was same. From pilot scale studies, the highest algal-bacterial biofilm growth rate after 3 weeks in one of the ponds at 5 cm depth was 3.63 gm^{-2} d^{-1}. If 6% dry weight of biomass is nitrogen (Lai and Lam, 1997), then the algal uptake rate under light conditions is 0.22 g-N m^{-2} d^{-1}. Therefore, the maximum amount of ammonium taken up by algae under light conditions is 0.48 mg l^{-1} and 0.63 mg l^{-1} for experiments run during 6 and 8 hours respectively. These values were subtracted from the overall amount of ammonium removed during the entire experimental period. The remainder gives the minimum amount reduced by nitrification.

Analysis of results

For the purpose of description in the following texts, all slopes of ammonia used in calculations of the nitrification rates of all experiments will be referred to as initial slopes. The 'initial slopes' is defined under the results section. Regression analysis using the F-test (95% confidence interval, at 0.05 levels) was used to check if the initial slopes for different treatments were statistically different.

Results

General trends

The ammonia concentration in light experiments 1 and 2 decreased with time and became constant after six hours (Figure 5.1). The slopes until six hours were used to calculate the nitrification rates (initial slopes). In the light experiments 3 and 4, the decrease was more linear (Figure 5.1) so the slope for the whole experimental time was taken as the initial slope and used to calculate the nitrification rates. In the dark experiments 1 and 3, there was a general decrease of ammonia with time until the DO were 4.1 mg l^{-1} and 2.3 mg l^{-1} (Figure 5.2a and b). Hereafter, the ammonia concentration started to increase with time. The reasons for this are not clear but might be due to ammonia release, which could have already started at the beginning of the experiment and resulted in a less steep slope. The nitrification rates for dark experiments 1 and 3 were calculated basing on the initial slopes until when ammonia concentration started to increase (Figure 5.2a and b). Dark experiment 2 behaved like the light experiments 3 and 4 where the ammonia decrease with time was linear. The DO in this experiment was kept high by bubbling with air possibly accounting for the similarity. In this case, the whole slope was used to calculate the nitrification rates.

Table 5 shows the nitrification rates for the all runs; note that the nitrification rates under light conditions were corrected for algal uptake. Table 5 also shows statistical comparisons between nitrification rates under light, dim light and dark conditions. The ranges of DO in the various experiments are also given. From table 5, it was seen that there was general

Table 5, Summary of Biofilm nitrification rates (g-Nm^{-2}d^{-1}) for all the 4 different experiments

Experiment	Nitrification rates (g-Nm^{-2}d^{-1}) based on initial ammonia slopes			DO ranges corresponding to the initial slope of the ammonia curve (mgl^{-1})			Significance (95% confidence)
	Light after correction	Dark	Dim light	Light	Dark	Dim light	Light Vs dark
1	1.82±0.29	1.62±0.31	-	8.9 - 6.4	9.0 - 5.1	n.a.	No
2	1.47 ±0.04	1.12±0.07	-	9.5 - 6.7	8.2 -7.7	n.a.	No
3	1.16±0.03	1.34±0.25	-	9.4 - 6.3	9.4 - 4.1	n.a.	No
4	0.95±0.24	0.16±0.12	0.56±0.22	9.8 - 8.3	2.3 - 1.1	4.9 - 3.2	Yes between light and dark Yes between light and dim light Yes between dim light and dark

1- Light and dark conditions with pH, temperature and oxygen left to vary
2- Light and dark conditions with pH, temperature and oxygen kept constant
3- Light and dark conditions with pH and temperature kept constant and oxygen left to vary
4- Light, dim light and dark conditions

Discussion
Effect of different variables on nitrification rates
1. Light and dark conditions, oxygen, pH and temperature left to vary

The ammonium concentration dropped from 21 mg l^{-1} to 17 mg l^{-1} for the biofilms exposed to light while under dark conditions, it dropped from 21 to 19 mg l^{-1}. The ammonia concentration in the control experiment dropped slightly throughout the experimental period showing that ammonia loss by volatilization was minimal. After correction for algal uptake, the nitrification rate for light condition 1 was 1.82±0.29 g-Nm^{-2}d^{-1} while that of dark condition 1 was 1.62±0.31 g-Nm^{-2}d^{-1} respectively. These results are within the range of those obtained by Leu et al., (1998); Craggs et al., (2000); McLean et al., (2000) and Lydmark et al., (2007). The nitrification rate under light conditions was slightly higher than the lowest nitrification rate of 1.70 g-Nm^{-2}d^{-1} obtained by Salvetti et al., (2006) for moving-bed biofilm reactors operated with influent ammonia and DO of 12 mg l^{-1} and 7.9 mg l^{-1} respectively. Although Salvetti et al., (2006) also obtained very high nitrification rates of 10.40 g-Nm^{-2}d^{-1}; pure oxygen was bubbled in their system; for our studies, oxygen was naturally provided by algae. For this experiment, statistical test shows no significant difference in the nitrification rate under dark and light conditions (Table 5). This is in disagreement with Verdegem et al., (2005) who found higher nitrification rates under light than dark conditions. A direct comparison to their work was not possible since they did not report the oxygen concentrations in their studies. Light is important for oxygen production by algal photosynthesis but if the DO in the dark is high, the effect of light ceases to be important. In this experiment, the bulk DO corresponding to the initial slopes of the ammonia in both light and dark conditions decreased from 8.9 to 6.4 and 9.0 to 5.1 mgl^{-1} oxygen respectively (Table 5). The DO in the dark decreased but remained above 3 mg l^{-1}

even after 8 hours (Figure 5.3). Probably, the decrease of DO in bulk water did not significantly affect the availability of oxygen to the nitrifier population in the biofilm (only bulk water and not biofilm DO was measured).

The effect of oxygen can be demonstrated from calculations based on stoichiometry of nitrification where 4.57 g of oxygen is required to oxidize 1g-N. The theoretical oxygen consumption was calculated for experiment 1 as follows: The volume of synthetic wastewater used in this experiment was 1.1 l and ammonium reductions under light and dark conditions based on the initial slopes were 4.4 mg NH_4-N and 2.2 mg NH_4-N respectively. If we assume that all the ammonium consumed was due to the nitrification process, 18.3 mg l^{-1} and 9.1 mg l^{-1} of oxygen was required for complete oxidation of ammonium under light and dark conditions. In reality, the DO decreased with only 2.5 mg l^{-1} (Table 5) under light conditions compared to the required 18.3 mg l^{-1} from the calculation. It is likely that the extra oxygen for the ammonia oxidation was produced by the photosynthesizing algal biofilm (Wolf et al., 2007). This is in agreement with our working hypothesis which suggested significant oxygen production in the algal biofilm. For the dark experiment, the decrease in DO was 3.9 mg l^{-1} during the initial experimental period (Table 5); this is less than 9.1 mg l^{-1} estimated from the calculations. Calculation of aeration by diffusion of oxygen from the atmosphere showed that it was insignificant hence the extra oxygen could not have been provided by this mechanism. There is a possibility that the extra oxygen was provided by DO already present in the biofilm before it was exposed to dark conditions. Alternatively, may be not all the ammonia reduction under dark experiment was due to nitrification. Probably some of the ammonia was taken up by bacteria for biomass development or there was an error in oxygen measurement at the end of the experiment. The results in the dark experiment showed that the initial oxygen concentration was sufficient to support nitrification.

From the above explanations, one of the mechanisms suggested to cause decrease of ammonia in both the light and dark experiments was ammonia oxidation. The other mechanism suggested was algal uptake. Nitrates accumulated during the experimental period showing that nitrification occurred (Figure 5.4). Although there was accumulation, the nitrate concentrations never exceeded 1 mg l^{-1}. Studies by Verdegem et al., (2005) suggest that algae prefer ammonium to nitrates as N-source. It is only when ammonium concentration is less than 0.03 mg l^{-1} Total Ammonia Nitrogen (TAN i.e. $NH_4 + NH_3$) that nitrite and nitrate uptake becomes important. Therefore nitrate uptake by algae could not explain the low nitrate concentrations observed during the experiment. The most probable explanation is denitrification since it has been found to occur in the deeper anoxic layers of the biofilms (Kuenen and Robertson, 1994; Revsbech et al., 2005). From the estimations of algal uptake (see under Material and Methods), it was seen that only 0.48 mg l^{-1} of NH_4-N was taken up by algae during the initial 6 hours of experiment 1 under light conditions. This was about 12% of the total NH_4-N removed during the light experiment. It is then suggested that algal uptake is the second principle removal mechanism after nitrification. Volatilization appeared to be negligible since the ammonia concentration in the control experiments was almost constant during the entire experimental period.

2. Light & dark conditions, oxygen and pH kept constant

In this experiment, the ammonia under light and dark conditions decreased from 19.2 to 10.9 mg l^{-1} and 12.4 mg l^{-1} respectively. The nitrification rates were 1.47±0.04 g-Nm^{-2}d^{-1} and 1.12±0.07 g-Nm^{-2}d^{-1} under light and dark conditions, respectively (Table 5). There was no significant difference (at 95% confidence interval) in nitrification rates and this could be explained by DO concentrations. The bulk water DO concentrations under light and dark conditions corresponding to the initial ammonia slopes were from 9.5-6.7 and 8.2-7.7 mg l^{-1} of oxygen respectively. The oxygen concentrations were kept high by bubbling with air and this seemed to favor nitrification, results which are similar to those of Goncalves and Oliviera, (1996). It appeared that keeping DO high under dark conditions had the same effect on nitrification as that of light and the associated oxygen production by photosynthesis. As long as there was sufficient oxygen in the bulk water, light intensity (as per this experiment) did not seem to affect the nitrification process.

3. Light & dark conditions, pH kept constant but oxygen left to vary

The nitrification rates for this experiment were 1.16±0.03 g-Nm^{-2}d^{-1} and 1.34±0.25 g-Nm^{-2}d^{-1} under light and dark conditions respectively. The difference was not statistically significant (Table 5). The minimum DO concentrations under light and dark conditions were 6.3 and 4.1 mg l^{-1}, so it seemed that the oxygen concentration under dark was still sufficient to support nitrification to a similar magnitude to that of light conditions. Even when the minimum DO in dark experiment 3 (4.1 mg l^{-1}) was lower than that of dark experiment 1 (5.1 mg l^{-1}, table 5), the nitrification rates were not significantly different. This implied that the DO concentration at this moment was still not limiting nitrification. These results are in agreement with Baskaran et al., (1992) who found minimal effect of light on nitrification as long bulk DO was still high enough to support the process.

4. Light, dim light & dark conditions

The nitrification rates for experiment 4 were 0.95±0.24, 0.56±0.22 and 0.16±0.12 g-Nm^{-2}d^{-1} for light, dim light and dark conditions, respectively. All the nitrification rates were significantly different from each other (Table 5). The mean biofilm biomass for the light, dim light and dark experiments were 18.7 ± 1.1 g VSSm^{-2}, 18.8± 5.8 g VSSm^{-2} and 15.0± 0.2 g VSSm^{-2} (n=2), respectively. The biomass of light and dim light conditions were similar hence cannot explain the difference between the two nitrification rates. In experiment 3, it was seen that under dark conditions at minimum DO of 4.1 mg l^{-1}, no significant difference was observed between the nitrification rate of light and dark; it was assumed that light did not have a direct effect on nitrification. However if the DO decreases to a minimum of 3.2 mg l^{-1} as in dim light experiment 4 (Table 5), a significant difference appeared. It seemed that at a certain point between 3.2 and 4.1 mg l^{-1} of bulk water DO under the given experimental conditions, nitrification became significantly reduced. In fact, further decrease of DO from a minimum of 3.2 mg l^{-1} under dim light to a minimum of 1.1 mg l^{-1} as under dark condition 4 significantly reduced nitrification rate. The limits of oxygen for effective biofilm nitrification under maximum COD 30-40 mg l^{-1} is proposed to be 2.5 mg l^{-1} (Chen et al., 1989) and it was uncertain if effective nitrification still occurred at COD above 30-40 mg l^{-1} (Baskaran *et al.*, 1992). Results from this study disagreed with those of Chen *et al.*, (1989) because the DO limit is higher i.e. between 3.2 and 4.1 mg l^{-1}. The difference can be explained by a higher COD of 100 mg l^{-1} of this study which increased competition for DO between heterotrophs and nitrifiers. The higher bulk water DO requirement could also be due to rate limiting diffusion of DO across the boundary layer of the biofilm. This study provides a new insight that for algae-bacterial biofilms, effective nitrification can still occur at COD levels of 100 mg l^{-1}.

Perspectives for design
Results from this study showed that high nitrification rates can be achieved in algal-bacterial biofilms under illuminated conditions. In tropical regions where there is sufficient sunlight during day time, designers should consider installing baffles in wastewater stabilization ponds to improve the nitrification capacity. This could especially benefit ponds that are highly loaded that the bulk DO even under sunlight is low (Kayombo *et al.*, 2002). Under those conditions, algal biofilms could improve nitrification. One of the design decisions to make is how deep the baffles should extend into the pond. It is obvious that extending the baffles into the anaerobic zone does not improve nitrification. This research could be used to estimate to what pond depth the baffles are useful; a depth until below the photic zone, until the point where the bulk water DO is between 3.2 and 4.1 mg l^{-1}. This depth could be appropriate for nitrification, since deeper than that point the nitrification rates will decrease.

Conclusions
- This study investigated indirect effect of light intensity, via the oxygen availability for the nitrifiers in the biofilm. This effect is absent when there is sufficient oxygen in the bulk liquid, but at bulk liquid DO values between 3.2 and 4.1 mg l^{-1}, biofilm nitrification rates are significantly reduced. The rates are even further reduced when the bulk oxygen concentration decreases below 1.1 mg l^{-1}.
- The results from this study demonstrated that the simple methodology used can be applied to investigate the effects of DO on algal-bacterial biofilm nitrification rates.

Acknowledgements
We are grateful for the financial support from the Netherlands government through Netherlands Fellowship Program. We also appreciate the financial assistance from the EU-Switch project contract 018530. The authors are also thankful to UNESCO-IHE laboratory staff for their assistance and support in the laboratory work.

References
APHA. (1995). Standard Methods for Examination of Water and Wastewater 19th Ed., Washington, D.C

Babu, M.A., Mushi, M.M., van der Steen, N.P., Hooijmans, C.M., Gijzen, H.J. (2007). Nitrification in Bulk Water and Biofilms of Algae Wastewater Stabilization Ponds. *Wat. Sci Tech.* **55** *(11) 93-101*

Baskaran, K., Scott, P.H., Connor, M.A. (1992). Biofilms as an aid to nitrogen removal in sewage treatment lagoons. *Wat .Sci. Tech. 26(7-8), 1707-1716*

Chen, G.H., Ozaki, H., Terashima, Y. (1989). Modelling of the simultaneous removal of organic substances and nitrogen in a biofilm. *Wat. Sci. Tech. 21(8-9), 791-804*

Craggs, L.J., Tanner, C.C., Sukias, J.P.S., Davies, C.R.J. (2000). Nitrification potential of attached biofilms in dairy wastewater stabilization ponds. *Wat. Sci. Tech. 42(10-11), 195-202*

Goncalves, R.F and Oliviera, F.F. (1996). Improving the quality of the effluent of the facultative stabilization pond by means of aerated submerged biofilters. *Wat. Sci. Tech. 33(3), 145-152*

Johnson, M and Mara, D.D. (2005). Aerated rock filters for enhanced nitrogen and faecal coliform removal from facultative waste stabilization pond effluent. *Wat. Sci. Tech. 51(12), 99-102*

Kayombo, S., Mbwettte, T., Mayo, A.W., Katima, J.H.Y., Jorensen, S.E. (2002). Diurnal cycles of variation of physical-chemical parameters in waste stabilization ponds. *Ecological Engineering, 18, 287-291*

Kuenen, G.J., and Robertson, L.A. (1994). Combined nitrification-denitrification processes. *FEMS Microbiology reviews 15, 109-117*

Lai, P.C.C and Lam, P.K.S. (1997). Major Pathways for Nitrogen Removal in Wastewater Stabilization Ponds. *Water, Air and Soil pollution, 94, 125-136*

Leu, H.G., Lee, C.D., Ouyang, C.F., Tseng, H. (1998). Effects of organic matter conversion rates of nitrogenous compounds in a channel reactor under various flow conditions. *Wat. Res. Vol. No. 3, 891-899*

Lydmark, P., Almstrand, R., Samuelsson, C., Mattson, A., Sorrensson, F., Lingren, P.E., Hermansson, M. (2007). Effects of environmental conditions on the nitrifying population dynamics in a pilot wastewater treatment plant. *Environmental Microbiology 9 (9), 2220-2233*

McLean, B.M., Baskaran, K., Connor, M.A. (2000). The use of algal-bacterial biofilms to enhance nitrification rates in lagoons: Experience under laboratory and pilot scale conditions. *Wat .Sci. Tech. 42 (10-11), 187-194*

Revsbech, N.R., Jacobsen, J.P., Nielsen, L.P. (2005). Nitrogen transformations in microenvironments of river beds and riparian zones. *Ecological Engineering, 24, 447-455*

Roeselers, G., van Loosdrecht, M.C.M., Muyzer, G. (2008). Phototrophic biofilms and their potential applications. *J Appl Phycol 20, 227-235*

Salvetti, R., Azzelino, A., Cinziani, R., Bonomo, L. (2006). Effects of temperature on tertiary nitrification in moving bed-bed biofilm reactors. *Wat. Res Vol.40, 2981-2993*

Verdegem, M.C.J., Eding, E.H., Sereti, V., Munubi, R.N., Santa-Reyes, R.A., van Dam, A.A. (2005). Similarities between Microbial and Periphytic Biofilms in Aquaculture Systems. In: Azim, M.E. (Ed), Periphyton, Ecology, Exploitation and Management. CABI International Wallingford, Oxford shire, UK, pp 311

Wolf, G., Picioreanu, C., van Loosdrecht, M.C.M. (2007). Kinetic Modeling of Phototrophic Biofilms: The PHOBIA Model. *Biotechnology and Bioengineering, vol. 97 No. 5, 1064-1079*

Xia, S., Lin, J., Wang, R. (2008). Nitrogen removal performance and microbial community structure dynamics response to carbon nitrogen ratio in a compact suspended carrier biofilm reactor. *Ecological Engineering, 32, 256-262*

Zimmo, O.R., van der Steen, N.P., Gijzen, H.J. (2004). Quantification of nitrification and denitrification rates in algal and duckweed based wastewater treatment systems. *Environmental Technology, 25, 273-282*

Chapter 6
Effect of operational conditions on the nitrogen removal in a pilot scale baffled wastewater stabilization ponds under tropical conditions

Chapter 6

Effect of operational conditions on the nitrogen removal in a pilot scale baffled wastewater stabilization ponds under tropical conditions

Abstract

Four pilot scale wastewater stabilization ponds (WSP) were set up in Kampala – Uganda and operated under low (period 1, 0.0057 g NH_4-N $l^{-1}d^{-1}$) and high (period 2, 0.0084 g NH_4-N $l^{-1}d^{-1}$) ammonia loadings. Pond 1 was operated as control while ponds 2, 3 and 4 were fitted with baffles having the same surface area for biofilm attachment but different configurations to induce different flow patterns. The major aim of this study was to investigate the performance in nitrogen removal of these ponds under the different operational conditions. The results of period 1 showed that the control pond performed better in nitrogen removal than the baffled ponds. This was probably due to the effect of TSS on light penetration. The TSS during period 1 was significantly higher than period 2. It is likely that the algae growing in the upper most layers blocked light penetration hence affecting the development of algal biomass and nitrifiers in the deeper parts of the baffles. In addition, TSS in effluents of WSP's is usually algal material so the higher TSS in pond 1 could mean more ammonia uptake by algae. In ponds 1 and 3, ammonia removal was positively related to effluent pH and organic nitrogen; increase in these variables in WSPs is linked to photosynthesis. So transformation of ammonia to organic nitrogen was important in all ponds during period 1. In period 2, the baffled ponds performed better than the control pond. Pond 3 performed best, followed by ponds 2, 4 and 1 which removed the least nitrogen. It is believed that the extra surface area for attachment of nitrifiers provided by baffles caused this observation. During this period, the effluent TSS was significantly lower and this could have improved the conditions on the baffles for nitrogen removal. Day-time oxygen concentrations in the water column during period 2 were higher which could have positively influenced nitrification-denitrification processes. When the two periods were compared, the nitrogen removal efficiencies during period 2 was higher than in period 1 in ponds 2, 3 and 4. It is believed that the higher influent ammonia, lower BOD and higher aerobic surface area of baffles during period 2 could have played a role; regression models of both periods showed that increase in nitrogen removal was correlated with increased influent ammonia. The nitrogen removal efficiency of pond 1 reduced during period 2 possibly due to lack of extra attachment surface area for nitrifiers in the ponds

Key words: Stabilization ponds; biofilm; ammonia removal; sewage; tropical conditions

Introduction

This study focused on the use of wastewater stabilization ponds in wastewater treatment. Wastewater stabilization ponds are advantageous to developing countries due to their simplicity, cost effectiveness, easy operation and use of solar energy (Veenstra and Alerts, 1996). Solar energy is abundant for most periods of the year in tropical countries (Mara, 1997; 2004); therefore this favors the use of wastewater stabilization ponds. However, their performance in the removal of nutrients such as nitrogen and phosphorous is less clear (Mergaert *et al.*, 1992; Lettinga *et al.*, 1993). Limitation of nitrogen removal in particular, has been associated with a narrow aerobic zone for nitrification (Baskran *et al.*, 1992) and lack of attachment surface for nitrifiers (Craggs *et al.*, 2000). Nitrifiers are also slow growers as compared to heterotrophic bacteria. In instances of high organic loading, they are more likely to be outcompeted by the heterotrophs. Increasing the aerobic zone, providing attachment surface and reducing competition from heterotrophs are possible approaches of favoring nitrifier growth.

In this study, the performance of wastewater stabilization ponds incorporated with baffles as attachment surface for nitrifiers was evaluated. The ponds were operated under two conditions and comparisons in ammonia removal between the two operational conditions were made. This chapter describes the performance of the ponds in terms of nitrogen removal efficiencies and statistical analysis; nitrogen mass balances were assessed in a subsequent study (Chapter 7).

Methodology
Description and operation of the pilot scale system

The pilot scale wastewater stabilization ponds are shown and described in chapter 2 (Figure 2). The operational conditions were also described in chapter 2.All the physico-chemical parameters were analyzed according to APHA, (1995).

Statistical analysis

Statistical t-tests and multiple regressions using SPSS® and Fowler and Cohen, (2003) were used in data analysis. Normal distribution was tested and data that did not satisfy the condition of normal distribution after log transformation were tested using non-parametric tests. Multiple regressions with categorical predictors (Shield, 2005) were done to compare the performance of ammonia removal between the ponds during the same period.

The goodness of the fits from multiple regressions was assessed using statistical diagnostics such as checking for influential cases that may bias the model. Influential cases are those that exert undue influence over the parameters in the model. Sometimes few influential cases bias the regression model (Shield, 2005). SPSS calculates the outcome of the model (dependent variable) with or without a particular case and compares the outcome. If the outcome did not change, then that particular case did not have undue influence on the model. There are several methods used to check for influential cases but the one used here is the Cook's distance. Cook's distance considers the effect of a single case on the whole model. If Cook's values of greater than 1 are obtained, then influential cases may be of concern.

Results

The results for period 1 and period 2 are presented and discussed separately and comparisons between the two operational conditions are made. The results are presented as means ± standard deviations. Therefore, the error bars on the graphs show standard deviations. In some cases, results have been presented as median ± standard deviation. Ammonia removal in this study is defined as the ammonia loss when influent and effluent ammonia concentrations are compared. Ammonia loss can be due to transformation to organic nitrogen by algae which settle or are washed out as TSS in the effluent. The ammonia can also be lost from the ponds through volatilization or transformed to oxidized forms by nitrification and permanently removed through denitrification. The nitrogen mass balances are presented in chapter 7.

Period 1

During period 1, the influent and effluent ammonia of the facultative pond were 79±9.0 and 34.2±6.9 mg l^{-1} respectively, 57% of the ammonia was removed. Therefore, the maturation ponds received an influent ammonia concentration of 34.2±6.9 mg l^{-1}. The average ammonia removals during period 1 for the maturation pond 1, 2, 3 and 4 were 21.2±4.4, 20.8±4.2, 20.7±5.2 and 9.6±5.1 mg l^{-1} respectively (Figure 6). Multiple regressions using ammonia removal and pond type were performed to test for the effect of pond type on ammonia removal. First, pond 1 was compared to ponds 2, 3 and 4 (Table 6). The resulting model could significantly explain 49% (F= 48.5; at p<0.001) of the variance of ammonia removal between the ponds. The results showed that the ammonia removal of pond 1 significantly differed from all the other ponds (Table 6). The negative β-values in table 6 during the first run showed that ammonia removal was significantly less in ponds 2, 3 and 4 in comparison to pond 1.

Table 6 Results for multiple regressions of ammonia removal and the pond type during period 1

Run	Variable	t-value	Sig (p)	Standardized coefficient (β)
1	Pond 1 versus pond 2	-3.2	<0.001	-0.231
	Pond 1 versus pond 3	-3.3	0.020	-0.235
	Pond 1 versus pond 4	-11.5	<0.001	-0.830
2	Pond 2 versus pond 3	0.05	0.957	-0.004
	Pond 2 versus pond 4	-8.4	<0.001	-0.598
	Pond 3 versus pond 4	-8.3	<0.001	-0.594

If p<0.05, then significant difference

Table 6.1Calculated and measured Kjeldahl nitrogen during period 1 and 2

Period	Pond	Measured effluent NH_4-N (mg-N l^{-1})	N-uptake by algae (as 6% effluent TSS) (mg-N l^{-1})	Calculated Kjeldahl-N (mg-N l^{-1})	Measured Kjeldahl-N (mg-N l^{-1})
1	1	9.5	17.1	26.6	18.6
	2	13.4	12.3	25.7	22.8
	3	13.5	14.3	27.8	23.5
	4	24.5	9.6	34.1	32.6
2	1	28.7	3.2	31.9	33.3
	2	15.2	2.8	18.0	18.8
	3	12.6	1.7	14.3	16.5
	4	19.1	1.7	20.8	22.1

Table 6.2 Results for multiple regressions of ammonia removal and the pond type during period 2

Run	Variable	t-value	Sig (p)	Standardized coefficient (β)
1	Pond 1 versus pond 2	11.8	<0.001	0.703
	Pond 1 versus pond 3	13.9	<0.001	0.832
	Pond 1 versus pond 4	8.7	<0.001	0.521
2	Pond 2 versus pond 3	2.2	0.032	0.129
	Pond 2 versus pond 4	-3.1	0.003	-0.182
	Pond 3 versus pond 4	-5.2	<0.001	-0.311

If p<0.05, then significant difference

Further multiple regressions were done by comparing ammonia removal of pond 2 with ponds 3 and 4. Pond 3 was also compared to pond 4 (Table 6). The results indicated that ammonia removal between pond 2 and 3 did not differ significantly. Pond 4 significantly differed from pond 2 and 3; and the negative β-values indicated that its ammonia removal is less than that of pond 2 and 3. In summary, pond 1 removed more ammonia followed by pond 2 and 3; pond 4 removed the least amount of ammonia. Therefore under these conditions, the un-baffled pond performed better in ammonia removal than the baffled ones. The total nitrogen removal efficiency of pond 1, 2, 3 and 4 were 53% 48%, 47% and 32% respectively.

Additional statistical analyses were performed to determine factors that could explain the difference of ammonia removal in the ponds. Multiple regressions were conducted for each pond using ammonia removal (dependent variable) and the following independent variables: influent ammonia, influent Kjeldahl nitrogen, influent and effluent BOD, pH, temperature and effluent organic nitrogen. Effluent concentrations were assumed to be equal to average concentrations in the pond, assuming perfect mixing. Effluent organic nitrogen was included as a variable because it is a measure for ammonia uptake by algae in the pond. Data for twelve months (December 2007 to 2008) were used for analysis.

Table 6.3 Regression equations 1-4 for maturation ponds 1 - 4 in period 1 while 5-8 for maturation ponds 1 - 4 in period 2

Period 1
Pond 1

$$NH_4\,N_{Rem} = 1.5 + (0.64 * Inf\,NH_4N) + (-3.6 * Inf\,pH) + (4.0 * Eff\,pH) + (0.28 * Eff\,OrgN) \qquad (1)$$

Pond 2

$$NH_4\,N_{Rem} = 7.4 + (0.39 * Inf\,NH_4N) \qquad (2)$$

Pond 3

$$NH_4\,N_{Rem} = 23.8 + (0.51 * Inf\,NH_4N) + (1.4 * Inf\,Temp) + (-1.9 * Eff\,Temp) + (0.45 * Eff\,BOD)$$
$$+ (0.30 * Eff\,OrgN) \qquad (3)$$

Pond 4

$$NH_4\,N_{Rem} = -7.3 + (0.49 * Inf\,NH_4N) \qquad (4)$$

Period 2
Pond 1

$$NH_4\,N_{Rem} = -56.7 + (0.74 * Inf\,NH_4N) + (0.15 * Inf\,BOD) + (3.8 * Eff\,pH) + (1.3 * Eff\,NO_3N)$$
$$+ (-0.02 * Eff\,Alk) + (-0.02 * Eff\,TSS) + (0.49 * Eff\,OrgN)$$
$$+ (0.58 * Eff\,Temp) \qquad (5)$$

Pond 2

$$log\,NH_4\,N_{Rem} = -1.1 + (1.4 * log\,Inf\,NH_4N) + (0.51 * log\,Eff\,pH) + (-0.04 * log\,Eff\,TSS)$$
$$+ (-0.09 * log\,Eff\,BOD) + (0.170 * log\,Eff\,Oxy5cm) \qquad (6)$$

Pond 3

$$NH_4\,N_{Rem} = -39.3 + (1.28 * Inf\,NH_4N) + (-1.47 * Inf\,pH) + (2.59 * Eff\,pH) + (-1.33 * Oxy70cm)$$
$$+ (0.50 * Eff\,OrgN) \qquad (7)$$

Pond 4

$$NH_4\,N_{Rem} = -77.8 + (1.9 * Inf\,NH_4N) + (-0.39 * Inf\,KjN) + (6.4 * Eff\,pH) + (-0.50 * Eff\,BOD)$$
$$+ (0.80 * Eff\,OrgN) + (0.36 * Oxy5cm) \qquad (8)$$

For pond 1, influent ammonia, day time influent and effluent pH as well as effluent organic nitrogen were significantly correlated to ammonia removal (Table 6.3). For ponds 2 and 4, it was only influent ammonia which was significantly correlated to ammonia removal. The results for pond 3 showed that ammonia removal was significantly correlated to influent ammonia, influent temperature, effluent temperature, effluent organic nitrogen and effluent BOD (Table 6.3).

Period 2
The average ammonia concentration entering and leaving the facultative pond was 75 ± 5 and 51.2 ± 4 mg l⁻¹ respectively i.e. only 32% of ammonia was removed. A two sample independent t-test showed a significant difference between the means (t (98) =26.7, p<0.05). The mean effluent ammonia of maturation pond 1, 2, 3 and 4 were 28.7 ± 4, 15.2 ± 4, 12.6 ± 3 and 19.1 ± 8 mg l⁻¹ respectively. Statistical t-tests showed significant differences between the influent and effluent ammonia of all the maturation ponds. This implied that significant amounts of ammonia were removed by the maturation ponds. Similarly, the effluent ammonia concentrations in all ponds were significantly different from each other implying that the ponds behaved differently in ammonia removal.

(Wrigley and Toerien, 1990; Lai and Lam, 1997). Increase in pH values can also lead to ammonia loss by volatilization. However, ammonia loss by volatilization was negligible, results similar to those of Zimmo *et al.,* (2003). The pH values obtained were usually less than 8 (Figure 6.5) and this resulted in lower ammonia loss by volatilization. In some ponds, ammonia removal increased with an increase in effluent nitrates and a decrease in effluent alkalinity; an indication of nitrification. Results from regression also showed that decrease in effluent BOD increased ammonia removal.

Conclusions

During period 1, under conditions of lower ammonia and higher BOD loading, the un-baffled pond performed better in nitrogen removal than the baffled ones. During period 2, all the baffled ponds performed better in nitrogen removal than the un-baffled one. Effluent TSS during period 2 was significantly lower than during period 1. This could have allowed more light penetration hence the higher algal biomass and oxygen concentration at deeper parts of the pond biofilm. This led to improved conditions for ammonia oxidation during period 2. Additionally, the lower BOD concentration in the influent during period 2 could have favored growth of nitrifiers hence improving the ammonia removal. Among the baffled ponds during period 2, pond 3 performed best, followed by ponds 2, 4 and pond 1 removed the least nitrogen. This demonstrated the importance of baffle configuration on the removal processes. It implied that the baffle configuration of pond 3 could be included in wastewater treatment designs for improving nitrogen removal in wastewater stabilization ponds.

Acknowledgements

We are grateful for the financial support provided by the Netherlands government through Netherlands Fellowship Program. We also appreciate the financial assistance from the EU-Switch project contract 018530. We are also thankful to the management and laboratory staff of Bugolobi Sewage Treatment Plant for their assistance and support in this research.

References

APHA (1995). Standard Methods for Examination of Water and Wastewater, 19th Ed., Washington, D.C

Baskaran, K., Scott, P.H. and Connor, M.A. (1992). Biofilms as an Aid to Nitrogen Removal in Sewage Treatment Lagoons. *Wat. Sci. Tech.* **26**(7-8), 1707-1716

Camargo Valero, M.A., Read, L.F., Mara, D.D., Newton, R.J., Curtis, T.P. and Davenport, R.J. (2009). Nitrification-denitrification in WSP: a mechanism for permanent nitrogen removal in maturation ponds. In: 8th IWA Specialist Group Conference on Waste Stabilization Ponds, 26 - 29 April 2009, Belo Horizonte/MG Brazil (Unpublished). IWA

Constable, J.D., Conor, M.A. Scott, P.H. (1989). The comparative importance of different nitrogen removal mechanisms in 5 west lagoon, Werribee treatment complex. 13th Australian Water and Wastewater Association Conference, Canberra

Craggs, L.J., Tanner, C.C., Sukias, J.P.S. and Davies, C.R.J. (2000). Nitrification potential of attached biofilms in dairy wastewater stabilization ponds. *Wat .Sci. Tech.42 (10-11) 195-202*

Fowler, J. and Cohen, L. (2003). Statistics for Ornithologists, BTO Guide No. 22

Gross, P.M., Scott, P.H., and Conor, M.A. (1994). Development of management procedure for maintaining nitrification in sewage treatment lagoons. Nutrient Removal from Wastewaters. Ed. N.J. Horan, P. Lowe, and E.I, Stentiford. Technomic Publishing, Lancaster. 47-54

Lai, P.C.C and Lam, P.K.S. (1997). Major pathways for nitrogen removal in wastewater stabilization ponds. *Water, Air and Soil pollution, 94: 125-136*

Lettinga, G., Man, A.D., ver der Last, A.R.M., Wiegant, W., van Knippenberg, K.,FrijnsJ., van Burren, J.L.C. (1993). Anaerobic treatment of domestic sewage and wastewater. *Wat .Sci. Tech. 27(9), 67-73*

Mara, D.D. (1997).Design Manual Waste Stabilization Ponds in India, Lagoon Technology International, Leeds-UK

Mara, D.D. (2004). Domestic wastewater treatment in developing countries. Earth scan, London

Mara, D.D., and Pearson, H.W. (1998). Design manual for waste stabilization ponds in Mediterranean countries. European Investment bank. Lagoon Technology International Ltd Leeds, England

McLean B.M., Baskran, K., and Connor, M.A. (2000). The use of algal-bacterial biofilms to enhance nitrification rates in lagoons: Experience under laboratory and pilot scale conditions. *Wat .Sci. Tech. 42(10-11), 187-194*

Mergaert, K., Vanderhaegen, B., Verstraete, W., (1992). Application and trends of pre-treatment of municipal wastewater. *Wat. Res. 26(10-11), 1025–1033*

Metcalf and Eddy (2003). Wastewater Engineering. Treatment and Reuse. Tchobanoglous, G., Burton, F.L., Stensel, H.D (Eds). 4th Ed. McGraw Hill, Inc., USA

Muttamara, S. and Puetpaiboon, U. (1997). Roles of Baffles in Waste Stabilization Ponds. *Wat. Sci. Tech. 35(8) 275-284*

Reed, S.C., Crites, R.W., Middlebrooks, E.J. (1995). Natural Systems for Wastewater Management and Treatment (2nd Ed). McGraw – Hill Inc

Shield, A. (2005). Discovering Statistics Using SPSS. Second Edition, Sage publication Ltd, London, SPSS Inc, (1997). SPSS® Base 7.5 syntax reference guides. SPSS Inc. p 814

Somiya, I. and Fujii, S. (1984). Material Balances of organics and nutrients in an oxidation pond. *Wat Res 18(3) 325-333*

Veenstra, S and Alaerts, G. (1996). Technology selection for pollution control. In: A. Balkema, H. Aalbers and E. Heijndermans (Eds.), Workshop on sustainable municipal waste water treatment systems, Leusdan, the Netherlands, 17-40.

Wrigley, T.J. and Toerien, D.F. (1990). Limnological aspects of small sewage ponds. *Wat. Res. Vol 24 No.1, 83-90*

Xia, X., Yang, Z. and Zhang, X. (2009). Effect of Suspended-Sediment Concentration on Nitrification in River Water: Importance of Suspended Sediment-Water Interface. *Environ. Sci. Technol. 43, 3681-3687*

Zanotelli, C.T., Medri, W., Belli Filho, P., Perdomo, C.C., Mullinari, M.R., and Costa, R.H.R. (2002). Performance of a baffled facultative pond treating piggery wastes. *Wat. Sci. Tech. 45(1) 49-53*

Zimmo, O.R., van der Steen, N.P. and Gijzen, H.J. (2003). Comparison of ammonia volatilization rates in algae and duckweed based wastewater stabilization ponds. *Wat. Res. 37, 4587-4594*

Zimmo, O.R., van der Steen, N.P., Gijzen, H.J. (2004). Quantification of nitrification and denitrification rates in algae and duckweed based wastewater treatment systems. *Env. Tech. 25 (3)273-282*

Chapter 7
Nitrogen mass balances for pilot scale biofilm stabilization ponds under tropical conditions

In Press as: M.A. Babu., N.P. van der Steen., C.M. Hooijmans., H.J. Gijzen. Nitrogen mass balances for pilot-scale biofilm stabilization ponds under tropical conditions. Bioresource Technology, doi:10.1016/j.biortech.2010.12.003

Chapter 7

Nitrogen mass balances for pilot-scale biofilm stabilization ponds under tropical conditions

Abstract

The main objective of this study was to model the mechanisms of nitrogen removal in biofilm wastewater stabilization pond based on simple nitrogen mass balance equations. Pilot scale biofilm maturation ponds were constructed at Bugolobi sewage treatment plant; Kampala – Uganda. The dimensions of the ponds were 4m x 1m by 0.8m depth; pond 1 served as control (without extra biofilm attachment surface) while in ponds 2, 3 and 4, fifteen baffles were installed in each. The control pond had a total surface area of 8 m^2 while the baffled ones had 23.2 m^2 each, including the pond wall. The baffles were arranged differently to induce different flow patterns. The ponds were operated under two loading conditions, i.e. period 1, 0.0057 g NH_4-N l^{-1} d^{-1} and period 2, 0.0084 g NH_4-N l^{-1} d^{-1}. Total nitrogen and TKN mass balances were made. Bulk water and biofilm nitrification rates were determined and used in the TKN mass balance. Results for total nitrogen mass balance showed that for both periods, denitrification was the major removal mechanism. Nitrogen uptake by algae was more important during period 1 than in period 2. The TKN mass balance predicted well effluent TKN for period 2 than period 1 when influent TKN/BOD ratio increased from 0.5 to 0.67. This could be due to fluctuations in algae density and ammonia uptake during period 1. No conclusions on reliability of mass balance model in period 1 were made. Under conditions as in period 2, the TKN mass balance model was a useful tool to predict performance of biofilm waste stabilization.

Key words: Biofilm, Bulk water, Stabilization ponds, Nitrification, Kjeldahl, Mass balances

Introduction

Nitrogen is causing eutrophication of water bodies worldwide and domestic wastewater is among the major sources of nitrogen disposal into the environment (de Godos *et al.*, 2010). The major effects of nitrogen pollution are manifested in eutrophication of surface waters as well as pollution of ground water. Algal blooms and proliferation of non desired aquatic weeds have been associated with nutrient pollution (Bolan *et al.*, 2009). Indirect effects of nitrogen pollution range from the blue baby syndrome to fish kills, from loss of aesthetic values of water to increased costs of water treatment. The possibility of nitrous greenhouse gas emissions from polluted water bodies has also been suggested (Gijzen and Mulder, 2001; Martinez *et al.*, 2009).

Many nations have adopted stringent effluent nitrogen standards, which are associated with increased treatment costs and demands upgrading of the already existing wastewater treatment (WWT) systems or construction of new systems. Simple and cost effective systems like wastewater stabilization ponds (WSP) in the standard configurations will in most cases not meet discharge standards. WSP research has shown large variations in performance for nitrogen removal, but it remains unclear which mechanisms are responsible for nitrogen removal under varying environmental conditions.

The principle routes of nitrogen transformation in wastewater stabilization ponds include nitrification, denitrification, sedimentation, volatilization, plant uptake and ammonification. In some studies, denitrification and sedimentation have been found to be the major removal mechanism (Zimmo *et al.*, 2004; Bolan *et al.*, 2009). Nitrification was previously thought not to occur in wastewater stabilization ponds due to low nitrate concentrations observed in the effluent (Pearson, 2005). However, McLean *et al.*, (2000), Wilkie and Mulbry, (2002), Zimmo *et al.*, (2004), Gonzalez *et al.*, (2008), de Godos *et al.*, (2010) and Molinuevo-Salces *et al.*, (2010) have shown nitrification to be an important process of nitrogen removal. Others (Aslan and Kapdan, 2006; Bolan *et al.*, 2009) have found ammonia volatilization to be a major nitrogen removal mechanism. Nitrogen uptake by algae has also been reported to be an important removal mechanism (Gonzalez *et al.*, 2008; Camargo Valero *et al.*, 2009; de Godos *et al.*, 2010); but upon death, sedimentation and mineralization of algae, internal cycling of ammonia occurs and this reduces the ammonia removal efficiency. All these mechanisms are an integral part of nitrogen removal mechanisms in wastewater stabilization ponds but which one prevails over the other depends much on the environmental and operational conditions (Zimmo *et al.*, 2004; Camargo Valero *et al.*, 2009; Gonzalez *et al.*, 2010).

Nitrification in WSPs could possibly be enhanced by providing attachment surface for slow growing nitrifiers (Zimmo *et al.*, 2004; Pearson, 2005; Gonzalez *et al.*, 2008) in biofilm waste stabilization ponds (BWSP). Biofilms have been found to improve nitrogen removal in wastewater stabilization ponds (McLean *et al.*, 2000; Craggs *et al.*, 2000 and Mara and Johnson, 2007). However, use of biofilms under tropical conditions and under different flow conditions has not been studied.

The aim of this study was to model the mechanisms of nitrogen removal in BWSP based on Total Kjeldahl Nitrogen (TKN) and total nitrogen (TN) mass balance equations. The nitrification rate in the TKN balance was based on measurement of nitrification rates in both the bulk water and biofilm in batch activity tests.

Methodology
Description of pilot scale system
The pilot scale wastewater stabilization ponds are shown and described in chapter 2 (Figure 2). The operational conditions were also described in chapter 2.All the physico-chemical parameters were analyzed according to APHA, (1995). This study was a continuation of the studies in chapter 6.

Nitrification batch activity tests
Nitrification rates in biofilms $(gm^{-2}d^{-1})$ were derived from the decrease in ammonia concentration with time in batch experiments. The decrease can be the result of volatilization, nitrification and algal uptake. Ammonia taken up by algae was measured (Babu et al., 2010) and volatilization was found to be negligible. Subsequently nitrification rates were calculated by subtracting the uptake by algae from the rate of ammonia decrease. Biofilms used in the tests were grown in the maturation ponds on biofilm plates of dimensions 3.0 by 8.0 cm which were vertically mounted on a frame and suspended in each maturation pond at 5 cm depth for more than 30 days. At the start of the batch experiments, the biofilm plates were rinsed with distilled water and biofilm nitrification batch activity tests performed as described by Babu et al., (2007, 2010).

Bulk water activity tests were performed using MP effluent, and the observed decrease in ammonia concentration was used to calculate the bulk water nitrification rates $(g\ l^{-1}d^{-1})$. Ammonia volatilization was found to be negligible and ammonia uptake by algae during the batch test was assumed to be the same as the average algae uptake rate in the ponds. Average algae uptake rate in the bulk water in the ponds was obtained by subtracting the influent organic nitrogen load $(g\ d^{-1})$ from that of the effluent and by dividing by the pond volume. The increase in organic nitrogen represents nitrogen taken up by algae because organic nitrogen is mainly algal material (Wrigley and Toerien, 1990; McLean et al., 2000; Camargo Valero et al., 2009).

Kjeldahl mass balance
Assumptions and limitations
The TKN mass balance (Equation 1) is a simplified model to describe TKN removal processes and could be used to estimate effluent TKN. The nitrification rate constants in Equation 1 were calculated from the nitrification rates at different pond depths, which were calculated based on maximum nitrification rates (obtained from the activity tests) and oxygen concentrations at depths of 5, 45 and 70 cm measured at 12.00 hours. That DO value was taken as the representative value though it is realized that in reality DO fluctuates strongly. Sensitivity analyses were done to check for the effect of this assumption on prediction of effluent TKN. The oxygen fluctuations take place over the duration of 24 hours, while the HRT is 6.4 days. Therefore it is expected that the effect of short term

fluctuations in oxygen on effluent nitrogen will be small. If this expectation is not true, one would need to measure oxygen profiles for all the four ponds on an hourly basis.

Kjeldahl mass balance equation

Nitrification rates of biofilm ($R_{biofilm}$) and bulk water ($R_{nitrbulk}$) obtained in the batch activity tests were used as parameters in the TKN mass balance (Equation 1) (adapted from Zimmo *et al.*, 2004) to predict the effluent TKN.

$$V_{reactor} \times \frac{d[TKN]}{dt} = Q \times [TKN]_{inf} - Q \times [TKN]_{eff} - (R_{Overall} \times V_{reactor}) + U_{vol} + S_{sed} \quad (1)$$

Where

$[TKN]_{inf}$	= Influent Kjeldahl Nitrogen (gm^{-3})
$[TKN]_{eff}$	= Effluent Kjeldahl Nitrogen (gm^{-3})
$R_{overall}$	= Overall nitrification rate ($gm^{-3}day^{-1}$), (($A_{surface} \sum R_{nitrbulk}(x) \Delta x)$ +
$(W \sum R_{biofilm}(x) \Delta x))/V_{reactor}$	
x	= depth below pond surface (m)
$R_{nitrbulk}$	= Nitrification rate in the bulk ($gm^{-3} d^{-1}$)
$A_{surface}$	= pond surface (m^2)
$R_{biofilm}$	= Nitrification rate in the biofilm ($gm^{-2}d^{-1}$)
W	= Total width of the biofilm (2 times total baffle width + total width of walls) (m)
Q	= Flow rate ($m^3 d^{-1}$)
U_{vol}	= Ammonia volatilization (gd^{-1})
S_{sed}	= TKN removal by sedimentation (gd^{-1})
$V_{reactor}$	= Total volume of pond (m^3)

All the terms in Equation (1) are expressed in gd^{-1}. The influent and effluent nitrogen flows as well as sedimentation and volatilization were converted to this unit and referred to as nitrogen fluxes (N-fluxes). Nitrification and denitrification rates were also changed to this unit and referred to as nitrogen conversions (N-conversions). Ammonia volatilization was calculated using Equation (2) developed by Zimmo *et al.*, (2004):

Ammonia volatilization rate ($g\text{-}N\ m^{-2}\ d^{-1}$) = $3.3[NH_3\text{-}N] + 4.90$ \quad (2)

Where $[NH_3\text{-}N]$ is calculated from Emerson *et al.*, (1975) as:

$$\% \text{ Unionised } NH_3 = \frac{100}{1+10^{(pKa-pH)}} \quad (3)$$

Where pKa is the ammonia dissociation constant

Table 7c Different parameters used in the mass balance model equation; the calculated and measured effluent TKN values are also given (Period 2).

Pond	Flow rate ($m^3 d^{-1}$)	Influent TKN (gm^{-3})	Influent TKN (gd^{-1})	Biofilm Fluxes (gd^{-1})	Bulk water fluxes (gd^{-1})	Volatilization rate (gd^{-1})	Sedimentation rate (gd^{-1})	Effluent TKN measured (gm^{-3})
1	0.526±0.01	59.1±5.5	31.1±2.9	5.3	3.5	0.040±0.03	1.06	33.3±4
2	0.525±0.01	59.1±5.5	31.0±2.9	15.3	3.9	0.037±0.02	1.22	18.8±4
3	0.519±0.01	59.1±5.5	30.7±2.8	19.3	3.5	0.033±0.02	1.21	16.5±3
4	0.523±0.01	59.1±5.5	30.9±2.9	12.3	6.0	0.026±0.01	1.18	22.1±8

113

Discussion
Nitrification rates
The biofilm nitrification rates obtained in the two periods (Table 7a) were within the range of other studies e.g. 0.48-0.72 gm^{-2} d^{-1} (Craggs et al., 2000), 0.72-0.96 gm^{-2} d^{-1} (McLean et al., 2000), 1.5-2.1 gm^{-2} d^{-1} (Babu et al., 2007) and 0.8 gm^{-2} d^{-1} (Lydmark et al., 2007). The bulk water nitrification rates in the two periods were in the range of 0.8 - 1.8 $gm^{-3}d^{-1}$, which were higher than 2.7×10^{-4} $gm^{-3}d^{-1}$ obtained by laboratory studies using exclusively synthetic wastewater (Babu et al., 2007). This could imply that field conditions using sewage favored more nitrifier growth in the bulk water.

Total Kjeldahl Nitrogen (TKN) and Total Nitrogen (TN) mass balances
The TKN mass balance (Tables 7b and 7c) showed that biofilm nitrification was the largest contributor to the mass balance, followed by bulk water nitrification, showing the potential of using biofilms to improve nitrogen removal. Volatilization was less than 1 gd^{-1} and this was in line with Zimmo et al., (2003) and Camargo Valero and Mara (2007) who found that this process was negligible for ammonia nitrogen removal in wastewater stabilization ponds; as long as pH remained below 8 (Zimmo et al., 2003). Sedimentation fluxes obtained were in the range of 0.8 to 2.4 g d^{-1} and these were within the values obtained by Zimmo et al., (2004). The sediments appeared dark green and this indicated that they were made up of mostly decayed algal material. Upon re-mineralization, the decayed algal matter releases ammonia in the water column. Very small quantities of sediments were collected and this could be an indication that most of it decomposed rather than accumulated. In terms of distribution, more sediment was collected at the inlet points of the ponds as compared to the middle and outlet positions.

The total nitrogen (TN) mass balance for period 1(Figure 7.1a) showed that denitrification accounted for 47%, 44%, 38% and 22% of the total influent nitrogen in ponds 1, 2, 3 and 4 respectively. Net algal uptake (effluent organic nitrogen) accounted for 18%, 19%, 20% and 17% of the total influent nitrogen in ponds 1, 2, 3 and 4, respectively. Denitrification is therefore a more important mechanism than algal uptake, except for pond 4. Note that denitrification was calculated as the difference of all terms in the mass balance. Therefore it represents not only denitrification per se but also any other term not included in the mass balance or errors. Caution should be taken when interpreting this term; it is an estimate of denitrification rather than true denitrification.

For period 2, the estimated denitrification in ponds 1, 2, 3 and 4 accounted for 36%, 59%, 65% and 56% of the total influent nitrogen (Figure 7.1b). Algal uptake accounted for 8%, 6%, 7% and 5% of the total influent nitrogen in ponds 1, 2, 3 and 4, respectively. Here the contribution of the estimated denitrification was also larger than algal uptake, and the difference is more pronounced than in period 1. Nitrification-denitrification being the most important removal mechanism is in agreement with results of Zimmo et al., (2004); Bolan et al., (2009) and Gonzalez et al., (2010). Sedimentation accounted for about 3-9% during both periods. Volatilization accounted for less than 1% of the total influent nitrogen hence can be considered to be negligible (Zimmo et al., 2003; Camargo Valero and Mara, 2007). The effluent organic nitrogen during period 1 was significantly higher than in period 2 i.e. during period 1, 9.1, 9.3, 10 and 8.1 mg-N l^{-1} while in period 2 they were 5.4, 3.6, 3.9 and 3.0 mg-N l^{-1} in ponds 1, 2, 3 and 4, respectively. Since effluent organic nitrogen in wastewater stabilization ponds is mostly contributed by algal material (Wrigley and Toerien, 1990), these results confirmed that algal uptake was a more important nitrogen removal mechanism during period 1. The amount of organic nitrogen that left the ponds in algal biomass can also be calculated as 6% of effluent TSS (Wilkie and Mulbry, 2002). The values thus obtained for period 1 for ponds 1, 2, 3 and 4 were 17.1, 12.3, 14.3 and 9.6 mg-N l^{-1}. Those for period 2 were 3.2, 2.8, 1.7 and 1.7 mg-N l^{-1}. The values during period 1 were significantly higher than period 2 which was consistent with the organic nitrogen results, indicating that estimation of organic nitrogen concentrations by TSS is reliable.

Using the TKN mass balance and nitrification rates to predict effluent TKN
The measured average effluent TKN for period 1 differed substantially from the predicted effluent TKN and had large standard deviations (Figure 7.2a). The model equations apparently did not predict the effluent values well under these unstable conditions, and this could be due to the large fluctuations in the algae density and ammonia uptake by algae. The mass balance discussed in the previous section showed that ammonia uptake by algae is an important mechanism during period 1. Another reason for the large variations in effluent TKN could indirectly be due to large fluctuations in influent BOD and TSS. The results for period 2 showed that the mass balance equation closely predicted the effluent TKN values (Figure 7.2b). The facultative pond was covered during this time; hence the maturation ponds received less TSS (due to limited algae growth in the FP). The BOD was also lower as compared to period 1. Both the influent BOD and TSS showed less variations (Figure 7.3a and 7.3b) and this could have been reflected in the effluent TKN; giving more accurate results. Under the experimental conditions, the TKN mass balance predicted better when the influent TKN/BOD ratio was 0.67 (period 2) than 0.5 (period 1). The TKN/BOD ratios are important in influencing the nitrifier population. For instance, the higher ratio during period 2 could have led to higher nitrification rates since growth of heterotrophic bacteria was less, primarily due to a lower organic loading (Downing and Nerenberg, 2008). Growth conditions for nitrifiers were therefore more favorable during period 2. And since the model is based on nitrification rates based on measurements under favorable TKN/BOD ratios, it is understandable that the model predicted better under the higher TKN/BOD ratios of period 2.

Despite the fact that only oxygen was taken into account during the calculations of N-conversions for nitrification and the other factors like pH and temperature were not considered, the model seemed to estimate the effluent TKN for period 2 relatively well. If the other factors would be considered, the model may still improve. More research on bulk water and biofilm nitrification rates at different depths, the effects of pH and temperature as well as an intensive campaign of measurements of oxygen profiles and other parameters during the whole day is recommended.

Sensitivity analyses
Since the model during period 2 could closely predict the measured effluent TKN, sensitivity analysis was performed for only this period. The results showed for all ponds a negligible effect of a 10% reduction in dissolved oxygen concentration, namely an increase of less than 1% effluent TKN. When oxygen concentration was reduced to 50% of its original value, the increase in effluent TKN was less than 1% for pond 1 and 2, while for ponds 3 and 4 it was 73 and 53 %, respectively. The ponds 3 and 4 had lower concentration of oxygen at 45 and 75cm so reduction of 50% of oxygen greatly affects effluent TKN. Sensitivity of the model to variation in oxygen

concentrations is therefore very limited for variations of 10% or less. The oxygen variations in the ponds usually occur over a time span of hours (Kayombo *et al.,* 2002), while the mean retention time of the ponds was 6.4 days. Short term changes in oxygen concentration are therefore less likely to affect the effluent TKN.

Conclusions

Results showed that biofilm nitrification (and subsequent denitrification) was the major pathway for TKN removal in BWSP, and more important than bulk water nitrification. Preference of nitrifiers for attached growth could explain these results.

TN mass balances showed that nitrification-denitrification, algal uptake and sedimentation were the principle nitrogen removal mechanisms in BWSP. The importance of nitrification-denitrification was more pronounced when the influent TKN/BOD ratio was increased.

The TKN mass balance model predicted effluent TKN better when influent TKN/BOD ratio increased from 0.5 to 0.67. Under the latter conditions, the model was a useful tool to predict performance of BWSP.

Acknowledgements

We are grateful for the financial support provided by the Netherlands Government through NUFFIC. We also appreciate financial assistance from the EU FP6 -SWITCH project - contract 018530. The authors are also thankful to the management and laboratory staff of Bugolobi Sewage Treatment Plant for their assistance and support in this research.

References

APHA. (1995). Standard Methods for Examination of Water and Wastewater 19th Ed., Washington, D.C.

Aslan, L.S and Kapdan, I.K. (2006). Batch kinetics of nitrogen and phosphorous removal from synthetic wastewater by algae. Ecol. Eng. **28**, 64-70.

Babu, M.A., Mushi, M.M., Steen N.P., Hooijmans, C.M., and Gijzen, H.J. (2007). Nitrification in Bulk Water and Biofilms of Algae Wastewater Stabilization Ponds, Wat. Sci. Tech. **55** (11) 93-101

Babu, M.A., Hes, E.M.A., van der Steen, N.P., Hooijmans, C.M and Gijzen, H.J. (2010). Nitrification rates of algal-bacterial biofilms in wastewater stabilization ponds under light and dark conditions. Ecol. Eng. 36, 1741-1746.

Bolan, N.S; Lauranson, S; Luo, J and Sukias, J. (2009). Integrated treatment of farm effluents in New Zealand's diary operations. Bioresource Technology **100**, 5490-5497.

Camargo Valero, M.A and Mara, D.D. (2007). Nitrogen Removal via Ammonia Volatilization in Maturation Ponds. Wat. Sci. and Tech, **55**(11), 87-92.

Camargo Valero, M.A and Mara, D.D and Newton, R.J. (2009). Nitrogen Removal in Maturation WSP Ponds via Biological Uptake and Sedimentation of Dead Biomass. Proceedings of 8th specialist wastewater stabilization pond conference, Belo Horizonte- Brazil.

Craggs, L.J., Tanner, C.C., Sukias, J.P.S and Davies, C.R.J. (2000). Nitrification   Potential of Attached Biofilms in Dairy Wastewater Stabilization Ponds, Wat. Sci. Tech. **42**(10-11), 195-202.

de Godos, I; Vargas, V.A; Blanco, S; Garcia-Gonzalez, M.C; Soto, R; Garcia-Encina, P.A; Becares, E. and Munoz, R. (2010). A comparative evaluation of micro algae for degradation of piggery wastewater under photosynthetic oxygenation. Bioresource Technology **101**, 5150-5158.

Downing, L.S and Nereberg, R. (2008). Effect of bulk liquid BOD concentration on activity and microbial community structure of a nitrifying, membrane-aerated biofilm. Appl. Microbiol. Biotechnol **81**, 153-162.

Emerson, K., Russo, R.E., Lund, R.E., Thurston, R.V. (1975). Aqueous Ammonia Equilibrium Calculations: Effect of pH and Temperature. Jour. Fisheries Res. Board of Canada **32** (12) 2379-2383.

Gijzen, H.J and Mulder, A. (2001). The Nitrogen Cycle Out of Balance. Water21, August 2001, 38-40.

Gonzalez, C; Marciniak, J; Villaverde, S; Garcia-Encina, P.A and Munoz, R. (2008). Microalgae-based processes for biodegradation of pretreated piggery wastewater. Appl. Microbiol. Biotech **80**, 891-898

Gonzalez, C., Molinuevo-Salces, B; Garcia-Gonzalez, M.C. (2010). Nitrogen transformations under different conditions in open ponds by means of microalgae-bacteria consortium treating pig slurry. Doi:10.1016/j.biortech.2010.09.052

Henze, M., Gujer, W., Mino, T., and Loosdrecht, M. (2000). Activated Sludge Models ASMI, ASMI2, ASM2d and ASM3. Scientific and technical report No **9**. IWA publishing.
Kayombo, S., Mbwettte, T., Mayo, A.W., Katima, J.H.Y and Jorensen, S.E. (2002). Diurnal Cycles of Variation of Physical-Chemical Parameters in Waste Stabilization ponds. Ecological Engineering **18**, 287-291

Lydmark, P., Almstrand, R., Samuelsson, C., Mattson, A., Sorrensson, F., Lingren, P.E and Hermansson, M. (2007). Effects of Environmental Conditions on the Nitrifying Population Dynamics in a Pilot Wastewater Treatment Plant, Environmental Microbiology **9** (9) 2220-2233

Mara, D.D and Johnson. (2007). Waste stabilization ponds and rock filters: solutions for small communities. Wat. Sci. Tech. **55** (7), 103-107

Martinez, J; Dabert, P., Barrington, S and Burton, C. (2009). Livestock waste treatment systems for environmental quality, food safety, and sustainability. Bioresource Technology **100**, 5527-5536

McLean, B.M., Baskran, K., and Connor, M.A. (2000). The Use of Algal-bacterial Biofilms to Enhance Nitrification Rates in Lagoons: Experience Under Laboratory and Pilot Scale Conditions. Wat .Sci. Tech. **42** (10-11), 187-194

Molinuevo-Salces, B., Garcia-Gonzalez, M.C., Gonzalez, C. (2010). Performance comparison of two bioreactors configurations (open and closed to the atmosphere) treating anaerobically degraded swine slurry. Bioresource Technology **101**, 5144-5149

Pearson, H.W. (2005). Microbiology of Waste Stabilization Ponds. In: Pond Treatment Technology (Ed). Shilton, IWA publishing Alliance House London UK pp

Wilkie, A.C and Mulbry, W.W. (2002). Recovery of dairy manure nutrients by benthic freshwater algae. Bioresource Technology **84**, 81-91

Wrigley, T.J., and Toerien, D.F. (1990). Limnological Aspects of Small Sewage Ponds. Wat. Res. Vol **24** No.1, 83-90

Zimmo, O.R., Steen, N.P and Gijzen, H.J. (2004). Quantification of Nitrification and Denitrification Rates in Algal and Duckweed Based Wastewater Treatment Systems. Environmental Technology, **25,** 273-282

Zimmo, O.R., Steen, N.P and Gijzen, H.J. (2003). Comparison of Ammonia Volatilization Rates in Algae and Duckweed-based Stabilization Ponds Treating Domestic Wastewater. Wat. Res. **37,** 4587-4594.

Summary

Summary

Domestic wastewater is a source of nitrogen in environmental systems. Nitrogen is known to be a major pollutant to the aquatic system. It causes eutrophication which leads to excessive algal growth or growth of other undesired water weeds such as water hyacinth. This results in disruption of the oxygen balance, release of toxins, loss of biodiversity and increased costs of water treatment; if the water resource is used as a source for drinking water production.

Wastewater stabilization ponds are treatment technologies that have been adopted by many developing countries. This is due to being cheap to construct, operate and maintain than activated sludge systems. However, they suffer from high effluent total suspended solids concentration (TSS), short-circuiting, long hydraulic retention time and ineffectiveness in removing nutrients like nitrogen. The problem of nitrogen removal is attributed to low nitrifier biomass present in the water column. Several studies have shown that the introduction of attachment surface for nitrifiers in the ponds improves nitrogen removal. However, information on the introduction of baffles as attachment surface for nitrifiers under tropical conditions is scarce.

This study focused on the effects of incorporation of baffles in pilot scale wastewater stabilization ponds. The pilot scale ponds were constructed at Bugolobi Sewage Treatment Works (BSTW) in Kampala, Uganda, and operated under tropical conditions. Settled wastewater was pumped from the sedimentation tank of BSTW into a 10 m^3 plastic anaerobic tank (AT) having a retention time of 3 days. The wastewater was then fed continuously by gravity at a flow rate of 2.1m^3 per day into a facultative pond (FP). The effluent of the FP was fed into four pilot scale maturation ponds (MP) of length, width and water depth of 4m by 1m by 0.8m at flow rates of 0.5m^3 per day. The details of the design and operation of the pilot scale are presented in figure 2 chapter 2 of this thesis. Pond 1 was operated as control while in ponds 2, 3 and 4, fifteen baffles of the same surface area were installed. The baffles had different configurations (pond 2: parallel to the flow, pond 3 and 4: perpendicular to the flow) inducing different horizontal and vertical flow patterns. The ponds were operated for two periods i.e. under an influent BOD of 72±45 mgl^{-1} and ammonia of 34±7 mgl^{-1} (period 1) and an influent BOD of 29±9 and ammonia of 51±4 mgl^{-1} (period 2). Introduction of baffles in wastewater stabilization ponds can affect their ecology, hydraulic characteristics and performance. This was studied and presented in different chapters. Laboratory studies on bulk water and biofilm nitrification rates were conducted, to complement the pilot scale studies.

The results of this study showed that nitrogen removal from wastewater can be improved by addition of extra attachment surface for nitrifiers. Experiments discriminating biofilm and bulk nitrification rates showed that biofilm nitrification rates were more important than bulk water nitrification rates (Chapter 4). Further laboratory experiments also showed that biofilm nitrification rates are significantly reduced at bulk water oxygen concentration of less than 3.2 mg l^{-1} (Chapter 5). The results for the pilot scale wastewater stabilization ponds showed that during period 1, the control pond performed better than the ones that had extra attachment surface (Chapter 6). Under such conditions, it was found that the bulk water TSS was high and this prevented light penetration into the deeper parts of the ponds resulting in reduction of aerobic biofilm area that is required for nitrification (Chapter 2). The higher BOD during period 1 also favored the growth of heterotrophic bacteria compared to the nitrifiers. Nitrifiers are slow growers and under high BOD loading, they are outcompeted by heterotrophic bacteria. When the light conditions and corresponding algal activity of the FP were changed to increase the ammonia concentration in the influent to the maturation ponds (by covering it with a black plastic sheet) in period 2, the influent BOD of the maturation ponds decreased. Under these conditions, it was found that the baffled ponds 2, 3 and 4 showed better N-removal than the control pond 1. The removal efficiency was attributed to more attachment surface for nitrifiers in ponds 2, 3 and 4. The TSS of the maturation ponds also decreased, allowing more light penetration hence more aerobic biofilm area became available to the nitrifiers. Additionally, the lower influent BOD during period 2 favored the growth of nitrifiers in the ponds. Therefore under these conditions, nitrogen removal in the baffled ponds became better than in the control pond. Among the baffled ponds, pond 3 performed better than ponds 2 and 4. The mean HRT of pond 2 was 7.5 days while that of pond 3 was 9.2 days (Chapter 3), this could explain the differences in performance. However, the mean HRT for pond 4 was 9.9 days, higher than that of ponds 2 and 3; the reason why its performance was lower is unclear. It may be a result of differences in the hydraulic flow and oxygen balance. This study showed that introduction of baffles in wastewater stabilization ponds affects both the ecology and the hydraulic characteristics of the ponds. The algal and zooplankton distribution in the four ponds differed, but how this related to the nitrogen removal is still unclear. The tracer study showed that the flow patterns of ponds 1 and 2 were similar indicating that the baffle arrangement in pond 2 did not affect the flow pattern. However, the tracer curves for ponds 3 and 4 were different and these ponds had a higher mean hydraulic retention.

This study also showed that nitrogen uptake by algae is significant in wastewater stabilization ponds. However, internal nitrogen cycling is known to occur when the dead algae in sediments are decomposed resulting in a release of nitrogen. Some of the nitrogen incorporated in algal biomass leaves the ponds through washout. The pH in the ponds was mostly below 8 hence ammonia volatilization was probably negligible. Total nitrogen mass balances showed that nitrification-denitrification was the major nitrogen removal mechanism (Chapter 7).

Outlook
Management of nitrogen in WSP effluents can be aimed to achieve two objectives i.e. protecting water sources, or reuse of nitrogen in agriculture. For protection of water resources, the aim is to maximize N-removal and while for N-reuse in irrigation; the effluent concentration need to be tailored to the type of crop. In this study, the effluent concentration of total nitrogen of pond 1, 2, 3 and 4 during period 2 was 36, 22, 18 and 24 mg l^{-1}. About 15-24% of this amount was organic nitrogen, the rest was ammonium. The effluent of all the ponds did not meet the European Standards for disposal of total nitrogen (< 15 mg l^{-1}) to sensitive surface waters. The effluents met the permitted standards of 25mg l^{-1} filtered BOD (CEC, 1991). In order to maximize reuse of resources, the effluents from the ponds can be reused in irrigation. Municipal wastewater with 20-85 mg l^{-1} total nitrogen causes no soil acidification (as observed from synthetic fertilizers) and can increase productivity. The algae washed out in the effluents can add organic matter and nutrients to the soil (WHO, 2006). The limitation of the reuse of pond effluents in agriculture is the relatively higher TSS of >20 mg l^{-1} which may cause problems for drip irrigation. The study has shown that the ponds had a variety of algal forms; this could be a potential for animal feed production (Hosetti and Frost, 1995). The ammonia concentration in the pond effluent was still sufficient for land scape irrigation provided the microbial quality is satisfactory. In tropical

regions with high solar radiation, wastewater from oxidation ponds has become acceptable for irrigation (Hosetti and Patil, 1988).

In this study, several zooplanktons were identified; some forms like the rotifers are suitable food source for small planktonic stages of fish larvae. The rotifers have both high nutritional value and high daily rates of production. They are important sources of food for freshwater fish larvae Rotifers of *Branchionus* species are considered to be suitable as the first food for fish (Lubzens, 1987; Martinez and Dodson, 1992). Although it is doubtful whether the effluents of the ponds in this study can be used for aquaculture, presence of high nutritional fish food sources and diversity of algae give an insight into the potential exploitation of these ponds for aquaculture (Roche, 1995).

The results of this study showed that baffles can be incorporated in maturation ponds and improve nitrogen removal under conditions of high influent ammonia and low BOD loading. In this study, this was achieved by covering the facultative pond; the other option can be through construction of deeper anaerobic ponds. This is advantageous because it saves land and the anaerobic pond can be utilized for biogas production. The major disadvantage is the increased costs of excavating deeper ponds, the trade-off between cost of land and cost of construction should be considered. Introduction of baffles in wastewater stabilization ponds increases the construction costs but cheap materials like wooden plates could be effective. To avoid the problem of dead volume, baffle configuration of pond 2 can be used since this did not affect hydraulic conditions of the ponds (Chapter 3). For up-scaling purposes, light plastic material can be used as baffles and these can be easily suspended by floaters. This decreases the cost involved in baffle installation and gives the flexibility of moving baffles to desired spacing. The baffle area required per pond volume based on the performance of pond 3 was calculated as $5m^2m^{-3}$ with the aerobic depth of pond 3 of 0.48m. This means for every $1m^3$, 5 baffles of 1m by 0.5m will be required. For a maturation pond of 50 m by 1m by 100 m, the total surface area required will be $25,000m^2$. This implies that 250 baffles with baffle spacing of 0.2m will be installed widthwise across the pond and 100 rows lengthwise. In order to cater for a spacing of 0.2m between the rows, the pond length should be adjusted to 120 m. This study shows that this addition of attachment surfaces in wastewater stabilization ponds can improve the process of nitrogen removal. Wastewater stabilization pond designs incorporating baffles is therefore recommended.

In terms of other parameters, introduction of deeper oxygenated zone and deeper sunlight penetration may also help in pathogen and BOD removal. The flow through anoxic and aerobic zones created by baffles may be useful for both BOD and N removal (and P). The ultimate aim is to optimize conditions to reach overall treatment objectives for a range of treatment parameters. This study was limited to nitrogen removal, further research on the effect of baffles on pathogen and phosphorous removal is recommended.

References
Council of the European Communities (1991). Council Directive of 21 May 1991 concerning urban wastewater treatment (91/271/EEC). *Official Journal of the European Communities,* L135/40 (30 May)

EPA (2004). Guidelines for water reuse. EPA 645-R-04-108. U.S. Environmental Protection Agency, Washington, D.C

Hosetti, B.B and Frost, H. (1995). A review of the sustainable value of effluents and sludges from wastewater stabilization ponds. *Ecol. Eng.* **5**, *421-431*

Hosetti, B.B and Patil, H.S. (1988). Evaluation of catalase activity in relation to physico-chemical parameters in a polluted river. In: V.P. Agrawal and L.D. Chaturvedi (Eds). Threatened habitats, Society of Biosciences, 393-404.

Lubzens, E. (1987). Raising rotifers for use in aquaculture. *Hydrobiol.* **147**, *245-255.*

Martinez, R.R and Dodson,S.I. (1992). Culture of the rotifer *Branchionus calyciflorus Pallas*. *Aquaculture,* **105**, *191-199.*

Roche, K.F. (1995). Growth of the rotifer *Branchionus calyciflorus Pallas* in diary waste stabilization ponds. *Wat. Res.* **29** *(10), 2255-2260.*

WHO (2006). Guidelines for Safe Use of Wastewater, Excreta and Grey water. Volume II, Wastewater use in Agriculture

Samenvatting

Samenvatting

Huishoudelijk afvalwater is een bron van stikstof in ecosystemen. Stikstof staat bekend als één van de belangrijkste vervuilende stoffen in aquatische systemen, die eutrofiëring veroorzaakt, wat weer leidt tot overmatige groei van algen of groei van andere aquatische onkruiden zoals waterhyacint. Dit veroorzaakt een verstoring van het zuurstof evenwicht, het vrijkomen van toxische stoffen en een verlies aan biodiversiteit en een toename in kosten voor drinkwater bereiding als dat oppervlaktewater daarvoor wordt gebruikt.

Oxidatievijvers vormen een zuiveringstechnologie die in veel ontwikkelingslanden toegepast wordt. Dit komt doordat ze goedkoper zijn om aan te leggen, te bedrijven en te onderhouden dan een actief slib systeem. Echter, deze technologie kampt met een hoog gehalte aan zwevende stof (TSS), kortsluitingstromen, lange hydraulische verblijftijden en een lage efficiëntie voor de verwijdering van nutriënten zoals stikstof. Het probleem met betrekking tot stikstof verwijdering wordt toegeschreven aan een lage concentratie aan nitrificerende biomassa in de waterkolom. Verscheidene studies hebben laten zien dat het introduceren van aanhechtingsoppervlakt voor nitrificerende bacteriën in de vijver een verbetering in de stikstof verwijdering teweeg brengt. Echter, kennis over de introductie van schotten die als aanhechtingsoppervlak dienen onder tropische condities is schaars.

Deze studie onderzoekt in het bijzonder de effecten van de introductie van schotten in oxidatie vijvers op proefschaal. De proefvijvers zijn gemaakt op het terrein van de Bugolobi Sewage Treatment Works (BSTW) in Kampala, Oeganda, en werden bedreven onder tropische condities. Voorbezonken afvalwater werd vanuit de voorbezinker van BSTW in een plastic anaërobe tank (AT) gepompt, die een verblijftijd van 3 dagen had. Het afvalwater stroomde vervolgens continu met behulp van verval en met een debiet van 2.1 m^3 per dag in de facultatieve vijver (FP). Het effluent van de FP stroomde naar vier stabilisatievijvers (MP) met een lengte, breedte en diepte van 4 bij 1 bij 0.8 meter, met een debiet van 0.5 m^3 per dag. De details van het ontwerp en het bedrijven van de proefvijvers staan beschreven in Hoofdstuk 2 van dit proefschrift. Vijver 1 werd bedreven als controle, terwijl 15 schotten in de vijvers 2, 3 en 4, elk met dezelfde afmetingen, geplaatst werden op verschillende wijze (vijver 2: parallel aan de stromingsrichting, vijver 3 en 4: loodrecht op de stromingsrichting) zodat verschillende horizontale en verticale stromingspatronen veroorzaakt werden. De vijvers werden bedreven voor twee periodes, namelijk met een influent BOD van 72±45 mg l^{-1} en een ammonia concentratie van 34±7 mg l^{-1} (periode 1) en een influent BOD van 29±9 en een ammonia concentratie van 51±4 mg l^{-1} (periode 2). De introductie van schotten in de stabilisatievijvers kan invloed hebben op hun ecologie, hydraulische eigenschappen en prestaties. Dit is onderzocht en wordt gepresenteerd in de verschillende hoofdstukken van dit proefschrift. Laboratoriumonderzoek naar nitrificatiesnelheden in bulk water en biofilm zijn ook uitgevoerd, om de pilot-schaal studies aan te vullen.

De resultaten van deze studie laten zien dat stikstofverwijdering uit afvalwater verbeterd kan worden door het toevoegen van extra aanhechtingsoppervlakte voor nitificeerders. Experimenten die onderscheid maakten tussen nitrificatie snelheden in de biofilm en in de waterkolom lieten zien dat nitrificatie in de biofilm belangrijker was dan in de waterkolom (Hoofdstuk 4). Aanvullende laboratoriumexperimenten lieten bovendien zien dat nitrificatiesnelheden significant afnamen bij zuurstofconcentraties in de waterkolom lager dan 3.2 mg l^{-1} (Hoofdstuk 5). Resultaten voor de proefvijvers lieten zien dat tijdens periode 1 de controle vijver het beter deed dan de vijvers met extra aanhechtingsoppervlakte (Hoofdstuk 6). Onder die condities bleek de zwevende stof in de waterkolom hoger te zijn en ook bleek dat dit het doordringen van licht in de diepere lagen van de vijver voorkwam. Dit resulteerde in een verkleining van het oppervlak aan aërobe biofilm, welke een voorwaarde is voor nitrificatie (Hoofdstuk 2). De hogere BOD tijdens periode 1 stimuleerde ook de groei van heterotrofe bacteriën in vergelijking met de groei van nitrificeerders. Nitrificeerders groeien langzaam en bij een hoge BOD belasting worden ze verdreven door de competitie met heterotrofe bacteriën. Toen de beschikbaarheid van licht en de bijbehorende algen activiteit van de FP veranderd werden om de concentratie ammonia te laten toenemen in het influent voor de stabilisatievijvers, door het overkappen van de vijver met zwart plastic (periode 2), nam de concentratie BOD in het influent voor de stabilisatievijvers af. Onder deze condities werd gevonden dat de vijvers met schotten (vijvers 2, 3 en 4) een betere N-verwijdering bereikten dan de controle vijver 1. De verwijderingsefficiëntie werd toegeschreven aan een toename van aanhechtingsoppervlak voor nitrificeerders in de vijvers 2, 3 en 4. De zwevende stof in de stabilisatievijvers nam ook af, wat ervoor zorgde dat licht dieper in de waterkolom door kon dringen, met als gevolg dat meer aërobe biofilm beschikbaar kwam voor de nitrificeerders. Daarnaast is de lagere BZV influent waarde tijdens fase 2 van voordeel op de groei van nitrificeerders in de vijvers. Daarom is onder deze omstandigheden de stikstofverwijdering in de vijvers met schotten beter dan in de controle vijver. Van de vijvers met schotten presteerde vijver 3 beter dan de vijvers 2 en 4. De gemiddelde verblijftijd in vijver 2 was 7,5 dagen, terwijl dat in vijver 3 9,2 dagen was (Hoofdstuk 3), wat het verschil in prestatie kan verklaren. Echter, de gemiddelde verblijftijd voor de vijver 4 was 9,9 dagen, hoger dan dat van vijvers 2 en 3; de reden waarom de prestatie lager was, is onduidelijk. Het is wellicht een gevolg van verschillen in de hydraulische stroming en zuurstofbalans. Deze studie laat zien dat het introduceren van schotten in oxidatievijvers zowel de ecologie als het stromingspatroon van de vijvers beïnvloedt. De soorten algen en zoöplankton verschilden per vijver, maar hoe dit gerelateerd is aan de verwijdering van stikstof is nog onduidelijk. Een studie naar spreiding in verblijftijden, m.b.v een tracer, liet zien dat het stromingspatroon van de vijvers 1 en 2 gelijk waren, wat erop duidt dat de schotten in vijver 2 het stromingspatroon niet beïnvloedden. Echter, het stromingspatroon in de vijvers 3 en 4 waren anders en deze vijvers hadden dan ook een hoger gemiddelde hydraulische verblijftijd.

Deze studie heeft ook laten zien dat opname van stikstof door algen in oxidatie vijvers aanzienlijk is. Echter, een interne stikstof cyclus vindt plaats waarbij afgestorven algen in het sediment afgebroken worden, waarbij weer stikstof vrijkomt. Een deel van de stikstof die in algen biomassa opgenomen was, verlaat de vijver door middel van uitspoeling. De pH van de vijvers was meestal beneden een waarde van 8, en daarom was vervluchtiging van ammonia waarschijnlijk verwaarloosbaar. De massabalans voor totaal stikstof liet zien dat nitrificatie-denitrificatie het belangrijkste mechanisme was voor stikstofverwijdering (Hoofdstuk 7).

Outlook
Het beheer van stikstof in het effluent van WSP is gericht op bereiken van twee doelstellingen, namelijk de bescherming van waterbronnen en het hergebruik van stikstof in de landbouw. Voor de bescherming van de watervoorraden ligt de nadruk op het maximaliseren van de stikstofverwijdering, terwijl voor stikstofhergebruik de concentratie in het effluent zou moeten worden afgestemd op het soort gewas in geval van irrigatie. In deze studie bedroeg de effluent concentratie van totaal stikstof van de vijvers 1, 2, 3 en 4 gedurende de periode 2: 36, 22, 18 en

24 mg l^{-1}. Ongeveer 15-24% hiervan was organische stikstof, de rest was ammonium. De effluent concentratie van vijver 3 is 50% lager dan van vijver 1, de controle. Echter, het afvalwater van alle vijvers voldoet niet aan de Europese normen voor de lozing van stikstof totaal (<15 mg l^{-1}) op gevoelig oppervlaktewater. Het afvalwater voldoet wel aan de toegestane normen van 25 mg l^{-1} gefilterd BOD (CEC, 1991). Met het oog op het optimaliseren van hergebruik kan het effluent van de vijvers worden gebruikt voor irrigatie; huishoudelijk afvalwater met 20-85 mg l^{-1} totaal stikstof veroorzaakt geen bodemverzuring (zoals waargenomen bij synthetische meststoffen) en kan de productiviteit verhogen. De algen die met het effluent uitspoelen kunnen organisch materiaal en voedingsstoffen toe voegen aan de bodem (WHO, 2006). De beperking voor hergebruik in de landbouw is dat de TSS van het effluent in alle vijvers > 20 mg l^{-1} was. Dit kan problemen veroorzaken in geval van druppel irrigatie. De studie heeft laten zien dat er verschillende soorten algen in de vijvers groeiden, die potentieel voor dierenvoeder gebruikt zouden kunnen worden (Hosetti en Frost, 1995). De ammonia concentratie in het effluent was voldoende voor irrigatie, op voorwaarde dat de microbiële kwaliteit voldoende is. In tropische regio's met veel zonlicht, is effluent uit stabilisatievijvers volledig geaccepteerd voor irrigatie (Hosetti en Patil, 1988).

In deze studie werden verschillende soorten zoöplankton geïdentificeerd; sommige typen zoals rotiferen zijn geschikt als voedsel voor de plankton fase van vissenlarven. De rotiferen hebben zowel een hoge voedingswaarde als een hoge dagelijkse productie. Ze vormen een belangrijke voedselbron voor larven van zoetwatervis. Rotiferen van het soort *Branchionus* worden verondersteld geschikt te zijn als eerste voedsel voor vissen (Lubzens, 1987; Martinez en Dodson, 1992). Hoewel het twijfelachtig is of het effluent uit de vijvers van deze studie gebruikt kan worden voor aquacultuur, geeft de aanwezigheid van goede voedingsbronnen voor vis en de diversiteit aan algen een inzicht in het mogelijke gebruik van deze vijvers voor aquacultuur (Roche, 1995).

Uit de resultaten van deze studie is gebleken dat schotten kunnen worden opgenomen in de vijvers om de stikstofverwijdering te verbeteren, onder de voorwaarde van een hoge influent ammonia concentratie en een lage BOD belasting. In deze studie werd dit bereikt door het bedekken van de facultatieve vijver, een andere optie is het bouwen van anaërobe, diepere vijvers. Dit is voordelig omdat het land bespaart en omdat de anaërobe vijver gebruikt kan worden voor de productie van biogas. Het grote nadeel is de toegenomen kosten van het graven van een diepe vijver, waarbij de kosten van grond moeten worden vergeleken met de kosten van de bouw. Invoering van schotten in de stabilisatievijvers verhoogt de bouwkosten, maar goedkope materialen, zoals houten platen, kunnen effectief zijn. Om het probleem van ongebruikt volume te voorkomen kunnen schotten worden gebruikt zoals in vijver 2 omdat dit de hydraulische conditie in de vijver niet beïnvloed (Hoofdstuk 3). Voor opschalingdoeleinden kan een lichte kunststof worden gebruikt als schotmateriaal dat gemakkelijk kan worden geplaatst door middel van drijvers. Dit vermindert de kosten die nodig zijn voor de installatie en omdat de schotten op ieder gewenste afstand kunnen worden geplaatst is het systeem flexibel. Het oppervlak van de schotten dat nodig is per vijvervolume gebaseerd op de prestaties van de vijver 3 werd berekend op 5m^{2}m^{-3} uitgaande van een aërobe diepte van vijver 3 van 0.48m. Dit betekent dat voor elke 1m^3, 5 schotten van 1m bij 0,5m nodig zullen zijn. Voor een vijver van 50m bij 1m bij 100m is de totale oppervlakte die nodig is 25.000 m^2. Dit houdt in dat er 250 schotten met een afstand van 0,2 m ertussen zullen moeten worden geïnstalleerd in de breedte van de vijver en 100 rijen in de lengte. Om tegemoet te komen aan een afstand van 0,2 m tussen de rijen moet de vijverlengte worden vergroot tot 120 m. Deze studie toont aan dat deze toevoeging van aanhechtingsoppervlakken in stabilisatievijvers stikstofverwijdering kunnen verbeteren. Het is daarom aanbevolen om bij het ontwerp van een stabilisatievijver schotten op te nemen.

De invoering van een diepere zuurstofrijke zone o.a. door middel van het laten doordringen van zonlicht kan ook helpen bij pathogeen en BZV verwijdering. De stroming door anoxische en aërobe zones gecreëerd door schotten kunnen nuttig zijn voor zowel BZV en N-verwijdering (en P). Het uiteindelijke doel is om de condities te optimaliseren zodat voor verschillende parameters de behandelingsdoelstellingen kunnen worden bereikt. Deze studie was beperkt tot de stikstofverwijdering, verder onderzoek naar het effect van schotten op pathogenen- en fosfaat-verwijdering wordt aanbevolen

Referenties

Council of the European Communities (1991). Council Directive of 21 May 1991 concerning urban wastewater treatment (91/271/EEC). *Official Journal of the European Communities*, L135/40 (30 May)

EPA (2004). Guidelines for water reuse. EPA 645-R-04-108. U.S. Environmental Protection Agency, Washington, D.C

Hosetti, B.B and Frost, H. (1995). A review of the sustainable value of effluents and sludges from wastewater stabilization ponds. *Ecol. Eng.* **5**, *421-431*

Hosetti, B.B and Patil, H.S. (1988). Evaluation of catalase activity in relation to physico-chemical parameters in a polluted river. In: V.P. Agrawal and L.D. Chaturvedi (Eds). Threatened habitats, Society of Biosciences, 393-404.

Lubzens, E. (1987). Raising rotifers for use in aquaculture. *Hydrobiol.* **147**, *245-255.*

Martinez, R.R and Dodson,S.I. (1992). Culture of the rotifer *Branchionus calyciflorus Pallas*. *Aquaculture,* **105**, *191-199.*

Roche, K.F. (1995). Growth of the rotifer *Branchionus calyciflorus Pallas* in diary waste stabilization ponds. *Wat. Res.* **29** *(10), 2255-2260.*

WHO (2006). Guidelines for Safe Use of Wastewater, Excreta and Grey water. Volume II, Wastewater use in Agriculture

Curriculum vitae

Mohammed Babu was born on March 21st 1973 in Mbale, Uganda. He went to St. Peters College Tororo for secondary education and later joined Islamic University in Uganda, Mbale. He graduated with a degree in B.Sc Educ (Bot-Zoo/Chem) with honors in 1998. In 1999, he was awarded a fellowship by the Netherlands Government to study MSc in Environmental Science and Technology at UNESCO-IHE Institute for Water Education, Delft-The Netherlands. He specialized in aquatic system analysis and his thesis was on "Potential Removal of Zinc by *C. Papyrus* from Artificial Wetland". In 2005, he was awarded another scholarship by the Netherlands government to study a PhD at UNESCO-IHE Institute for Water Education under sandwich construction program. His PhD research was financed by EU-SWITCH project. Since graduation in 2001 to date, he has been lecturing in the department of environmental studies at the Islamic University in Uganda, Mbale.

His address in Uganda is; Islamic University in Uganda, P. O. Box 2555 Mbale, Uganda.
Telefax: +256454433502
Email: Babumohd@yahoo.com

129

T - #0115 - 071024 - C26 - 244/170/7 - PB - 9780415669467 - Gloss Lamination